Left Versus Right Asymmetries of Brain and Behaviour

Left Versus Right Asymmetries of Brain and Behaviour

Special Issue Editor

Lesley J. Rogers

MDPI • Basel • Beijing • Wuhan • Barcelona • Belgrade

MDPI

Special Issue Editor
Lesley J. Rogers
University of New England
Australia

Editorial Office
MDPI
St. Alban-Anlage 66
4052 Basel, Switzerland

This is a reprint of articles from the Special Issue published online in the open access journal *Symmetry* (ISSN 2073-8994) from 2018 to 2019 (available at: https://www.mdpi.com/journal/symmetry/special_issues/Left_Versus_Right_Asymmetries_of_Brain_and_Behaviour)

For citation purposes, cite each article independently as indicated on the article page online and as indicated below:

LastName, A.A.; LastName, B.B.; LastName, C.C. Article Title. *Journal Name* **Year**, *Article Number, Page Range.*

ISBN 978-3-03921-692-5 (Pbk)
ISBN 978-3-03921-693-2 (PDF)

Cover image courtesy of Gisela Kaplan.

Contents

About the Special Issue Editor

Lesley J. Rogers is a Fellow of the Australian Academy of Science and Emeritus Professor at the University of New England, Australia. After being awarded a First-Class Honours degree by the University of Adelaide, she studied at Harvard University in USA and then the University of Sussex, UK. She was awarded a Doctor of Philosophy and, later, a Doctor of Science from the University of Sussex, UK. After returning to Australia, she held academic positions at Monash University and the University of New England. Her publications, numbering over 500, include 18 books and over 280 scientific papers and book chapters, mainly on brain and behaviour. In the 1970s, her discovery of lateralized behaviour in chicks was one of three initial findings that established the field of brain lateralization in non-human animals, now a very active field of research. Initially, her research was concerned with the development of lateralization in the chick as a model species, and the importance of light stimulation before hatching, which she investigated at the neural and behavioural levels. She then compared lateralized behaviour in different species spanning from bees to primates and, more recently, has focused on the advantages of brain asymmetry and the link between social behaviour and population-level asymmetry. Her other roles include Editor of the journal *Laterality* and Academic Editor of numerous other scientific journals.

Preface to "Left Versus Right Asymmetries of Brain and Behaviour"

Asymmetry of the brain and of behaviour is a characteristic of a wide range of vertebrate species, as shown by an increasing number of studies testing animals in the laboratory and in the natural environment. Some asymmetries of behaviour have also been found in invertebrate species. Given its ubiquity, lateralization must confer an advantage for survival, despite the apparent disadvantages of side biases in perception and response. A disadvantage of lateralized responding is evidenced by the fact that many species are more likely to respond to a predator when it is seen on their left side and to their prey when it is seen on their right side. How do different species deal with these asymmetries? The topics covered in this book address this question and report further evidence of lateralized brain and behaviour in non-human species. In addition, the brain function involved in lateralized processing and control of response is discussed, and also the relationship between lateralized behaviour and animal welfare.

The paper by Frasnelli and Vallortigara addresses the question of why the majority of individuals in a population are lateralized in the same direction (population-level lateralization). They show that, although the cognitive advantage of having a lateralized brain places no constraints on the direction of lateralization, population-level lateralization develops as an evolutionary stable strategy when lateralized organisms must co-ordinate their behaviour with other lateralized organisms. This explains why population-level lateralization is a characteristic of social species. In this paper, the authors affirm that population-level asymmetry is also an advantage in so-called "solitary" species when individuals have to interact, as in aggressive and mating behaviour. They clarify an important point about inter-individual interaction and the evolution of lateralization as an evolutionary stable strategy.

The paper by Boeving and Nelson considers the link between social and affiliative behaviour from another perspective; by relating research showing that lateralization influences social structure in spider monkeys. Previous research had shown that social affiliative behaviour—embrace and face-embrace—in spider monkeys is left-side biased. In this paper, the authors apply social network analysis and find that laterality of affiliative behaviour influences social structure. Network patterns that are left-lateralized for affiliative behaviour are more cohesive than those that are right lateralized.

The paper by Üver, Xiao and Güntürkün reports research on the mechanism by which the brain deals with the conflicting responses elicited by each hemisphere's differing reaction to the same stimulus. In short, they reveal how one hemisphere achieves dominance (meta-control) over the other. Experiments addressing this issue involved sectioning the anterior commissure of pigeons, the largest commissure connecting the left and right sides of the avian brain. The results showed that meta-control is modified by interhemispheric transmission via this commissure, although it does not seem to depend entirely on it. The results suggest that the two hemispheres compete to take control of a particular behaviour and they do so on the basis of their processing speed. Since the hemisphere specialised to respond to a particular stimulus processes information faster than the other hemisphere, it takes control of the response.

From early research on lateralization of song production in the zebra finch, there has been speculation about the possibility that lateralization in this species differs from the general pattern found in other avian species and generally in vertebrates. The chapter by Rogers, Koboroff and Kaplan discusses more recent evidence refuting this idea and reports experimental evidence showing that population-level lateralization is present in preferred-eye use by zebra finches when they view a predator. Since zebra finches often alternate looking with the monocular field of one eye and then the other eye, a new method had to be developed in order to score eye preferences. The experiments showed that the birds have a significant preference to view a monitor lizard with their left-eye (using their right hemisphere). This result is discussed together with evidence of other asymmetries in zebra finches, for visual searching and courtship behaviour and for processing, producing and learning of song. The authors conclude that, contrary to earlier suggestions, the zebra finch brain is lateralized with the same pattern as that of that found in other vertebrate species.

Hausberger and colleagues consider lateralization of auditory processing. Auditory stimuli of differing salience (e.g., familiar versus novel sounds) were presented to Campbell's monkeys and only novel sounds elicited laterality. The monkeys had a significant right-hemisphere preference to attend to novel sounds but no preference to attend to familiar sounds. The authors also considered auditory lateralization in starlings. In starlings, the right hemisphere was found to process sounds of individual identity, whereas the left hemisphere was more involved in processing socially meaningless stimuli. The authors suggest an attention-based explanation to reconcile the different hypotheses about right-hemisphere specialisation.

Although many behavioural responses have a directional bias within the population, some types of laterality occur with equal numbers of left and right biased individuals in the population. Laterality in scale-eating cichlid fishes is such an example, discussed in the chapter by Hori and colleagues. These fish have asymmetry of the body, in the direction of the mouth opening either to the left or right side. The distribution of laterality within a population is bimodal (anti-symmetry). The authors have investigated the relationship between behavioural laterality and morphological asymmetry in two species studied over three decades. They found that the dimorphism is maintained dynamically with a cycle of four years oscillating between more left and more right individuals. This cycling is caused by frequency-dependent selection (the minority type having an advantage) between predator and prey species. Since both predator and prey fish are lateralized, the authors examine cross-predation versus parallel-predation in terms of the physical and sensory abilities of fishes.

The development of lateralization in Port Jackson sharks is dependent on temperature of the sea, as Pouca et al. report. They found that, under water temperatures predicted for the end of the century, development of sharks is affected, as seen by measuring preferences of direction taken during a detour test. Sharks incubated at the higher temperature had stronger lateralization (biased to detour to the right) than did sharks incubated at current sea temperature. The authors suggest that this change in lateralization might be a way by which the species could cope with deleterious effects of climate change.

Two papers deal with different aspects of laterality in dogs and its relationship to behaviour and welfare. The paper by Siniscalchi and colleagues reports on turning behaviour in sheepdogs. The dogs showed significantly more aggressive behaviour toward the sheep when they were circling the herd in an anticlockwise direction and so could see the sheep in their left visual field and process the information in their right hemisphere. Dominance of the right hemisphere in aggressive behaviour has been found also in a number of other vertebrate species. As the authors say, this relationship between motor lateralization and aggressive behaviour has practical implications for training sheepdogs.

The paper by Wells and colleagues relates laterality to the welfare of dogs. The subjects were rescued dogs and they were tested during the first week after they had been placed in a rescue shelter.

Paw preference measured in a food-retrieval task was linked to stress-related behaviour. The results showed that stronger left-paw preference was associated with higher stress-related behaviour, such as frequent change of state, vocalisations and lower body posture. This finding is in keeping with other findings of the association between left-limb preference and vulnerability to stress. The authors suggest that testing paw preference may be a useful tool for detecting different coping strategies in dogs entering a kennel environment and for targeting individuals at risk of experiencing reduced welfare.

Lesley J. Rogers
Special Issue Editor

symmetry

MDPI

Concept Paper

Individual-Level and Population-Level Lateralization: Two Sides of the Same Coin

Elisa Frasnelli [1,*] **and Giorgio Vallortigara** [2]

[1]   School of Life Sciences, University of Lincoln, Lincoln LN6 7DL, UK
[2]   Center for Mind/Brain Sciences, University of Trento, Piazza della Manifattura 1, I-38068 Rovereto, Italy;
      giorgio.vallortigara@unitn.it
*   Correspondence: efrasnelli@lincoln.ac.uk

Received: 21 November 2018; Accepted: 7 December 2018; Published: 11 December 2018

Abstract: Lateralization, i.e., the different functional roles played by the left and right sides of the brain, is expressed in two main ways: (1) in single individuals, regardless of a common direction (bias) in the population (also known as individual-level lateralization); or (2) in single individuals and in the same direction in most of them, so that the population is biased (also known as population-level lateralization). Indeed, lateralization often occurs at the population-level, with 60–90% of individuals showing the same direction (right or left) of bias, depending on species and tasks. It is usually maintained that lateralization can increase the brain's efficiency. However, this may explain individual-level lateralization, but not population-level lateralization, because individual brain efficiency is unrelated to the direction of the asymmetry in other individuals. From a theoretical point of view, a possible explanation for population-level lateralization is that it may reflect an evolutionarily stable strategy (ESS) that can develop when individually asymmetrical organisms are under specific selective pressures to coordinate their behavior with that of other asymmetrical organisms. This prediction has sometimes been misunderstood as it is equated with the idea that population-level lateralization should only be present in social species. However, population-level asymmetries have been observed in aggressive and mating displays in so-called "solitary" insects, suggesting that engagement in specific inter-individual interactions rather than "sociality" per se may promote population-level lateralization. Here, we clarify that the nature of inter-individual interaction can generate evolutionarily stable strategies of lateralization at the individual- or population-level, depending on ecological contexts, showing that individual-level and population-level lateralization should be considered as two aspects of the same continuum.

Keywords: lateralization; individual-level; population-level; evolution; ESS; social interactions

1. Introduction

Lateralization, defined as the different specialization of function of the left and right sides of the nervous system, is a widespread phenomenon in the animal kingdom. In the last three decades, many studies have provided evidence that many animal species, from the evolutionarily closest to the most evolutionarily distant from humans, show asymmetrical biases in behavior [1]. Examples range from the asymmetrical use of limbs to handle objects or perform motor activities (for a review, see [2]) to the asymmetrical use of sensory pair organs, such as eyes, nostrils, ears, and antennae to detect a specific stimulus, such as a potential predator; from motor biases in escape directions or navigation to the asymmetrical processes involving learning and memory and the processing of emotions [3,4]. All this evidence of brain and behavioral asymmetries in vertebrates [1], together with some in invertebrates [5,6], suggests that having an asymmetrical brain must confer advantages to complex brains, as well as to "simpler" ones.

Lateralization varies in strength (an individual may be less or more strongly lateralized) and direction (left or right) among individuals of the same species, of different species, and also depending on the task considered. Moreover, it can be present at the individual- or population-level (when most individuals within the population show the same direction of bias). Population-level lateralization has been explained as a consequence of selective social pressures that have pushed individuals to coordinate with each other and align their biases in the same direction [7]. In this paper, we discuss the advantages and disadvantages connected with having a (less or more) strong lateralized brain and the complexity of this fascinating phenomenon, while claiming that individual-level and population-level lateralization should be interpreted as two aspects of the same continuum.

2. Advantages of Having an Asymmetrical Brain (at the Individual Level)

Having an asymmetrical brain provides several advantages (see, for an extensive discussion, [7–9]). If the left and right sides of the brain perform different functions, it is possible to save energetic resources in cognitive tasks. Indeed, lateralization avoids the duplication of functions in the two hemispheres (otherwise, animals should probably have a brain double the size). Another big advantage related to lateralization consists of the possibility to separately and simultaneously process external stimuli, increasing the efficiency of the cerebral capacity. This is particularly easy to observe in animals with laterally placed eyes, such as birds, which mainly have monocular vision when using their lateral visual fields, i.e., they use their right and their left eye separately. More precisely, in birds, the lateral part of the right retina only communicates with the left hemisphere and vice versa. Because of this peculiarity, species such as the domestic chick *Gallus gallus* have been widely studied to assess the preferential use of the left and right side of the brain in specific tasks [10,11]. Chicks are better at discriminating grains of food from pebbles randomly mixed on the ground when they use their right eye (and thus their left hemisphere as, in vertebrates, the left hemisphere controls the right part of the body and vice versa; [12] see also for quails [13]). At the same time, chicks are better at detecting the presence of a potential predator when this appears in their left visual hemi-field (and it is perceived by their left eye and thus by the right hemisphere; [12]). Because of this functional specialization, chicks can feed from the ground using their right eye and, simultaneously, they can keep their left eye ready to respond to and protect themselves from potential predators [14].

Furthermore, when one hemisphere controls a specific behavior (for example, detecting potential predators), it is not competing with the other hemisphere to take control of that specific behavior. This leads to a more rapid and efficient response. Cerebral lateralization is indeed linked to better cognitive performances. Some studies have shown that more strongly lateralized individuals are more successful in some cognitive tasks compared to weakly lateralized conspecifics. In fact, behavioral asymmetries may vary not only in direction, but also in strength, among different individuals of the same species: some individuals can be more or less left-biased, others right-biased, and yet others unbiased. This is the case, for example, for chimpanzees, when fishing for termites using a stick: individuals with a strong preference to consistently use one hand (regardless of whether it is the left or the right one) are more efficient than individuals that do not have any preference to use one or the other hand [15]. Children with consistent early hand preferences exhibit advanced patterns of cognitive development compared to children who develop a hand preference later, although this could be a matter of synchronized development [16]. Strongly lateralized parrots showing a significant foot and eye preference are better at solving novel problems, such as a pebble-seed discrimination test and a string-pull problem, than less strongly lateralized parrots [17]. In domestic chicks (*Gallus gallus*), a right-eye superiority has been documented in inhibiting pecks at pebbles while searching for grain and this ability is impaired when lateralization is not present [14,18]. Similarly, pigeons (*Columba livia domestica*) with the strongest eye lateralization in discriminating grains from pebbles are the most successful in selecting grains when tested binocularly, suggesting that stronger lateralization increases the efficiency of a performance [19].

Surprisingly, insects also seem to have a preference for using one limb. Locusts crossing a gap have been shown to preferentially use the left or the right leg in this task [20]. Different individuals showed different biases not only in the direction (left or right), but also in the strength of the bias. However, as in chimpanzees [15], the individuals with a strong preference were those that made fewer mistakes in the task and thus were most successful [21]. This suggests that in this specific context, stronger lateralization confers a benefit in terms of improved motor control. Strong lateralization also seems to influence learning ability, as shown in larval antlions (*Myrmeleon bore*), with strong lateralized righting behavior being better at associating a vibrational cue with prey removal [22].

Not only behavioral asymmetries may vary in direction and strength among different individuals of the same species; biases can also change, depending on the task that an animal is performing (e.g., handedness in marmosets, [23]). This indicates that lateralization is a complex phenomenon that varies at the species, group, and individual level, bringing us to the question of what are the advantages of having individuals with different biases in the population. Individuals with a strong lateralization seem to have an advantage in terms of improved motor control [15,21] or problem solving [14,17–19]. However, in strongly lateralized fish, a consistent lateral bias to turn in one direction reduces their ability to orient in a maze [24]. This makes the scenario more complex and opens further questions about the optimal degree (and direction) of bias that an individual should have, depending on the task and functional context.

In sage grouse (*Centrocercus urophasianus*), successfully mating males are in general more strongly lateralized in courtship behavior than non-mating males, but this depends on the behavior of the male and the social environment in which he is acting [25]. Larger male fallow deer (*Dama dama*) display a greater tendency to show a right-sided bias when terminating the parallel walk during fights and they terminate parallel walks sooner than smaller individuals, suggesting that lateralization provides a mechanism by which contestants can resolve contests at a low cost [26]. Accordingly, in dyadic contests, domestic pigs (*Sus scrofa*) with strong lateralization in the orientation towards their opponent (regardless of the direction) have a shorter contest duration than conspecifics with a weak bias. However, although lateralization seems to play a role in conflict resolution, it does not influence fighting success, as winners and losers showed a similar strength and direction of bias [27]. Less lateralized wild elk (*Cervus canadensis*) for front-limb biases (i.e., handedness) respond more intensely to aversive stimuli (predator-resembling chases by humans), but the same animals are also more inclined to reduce their flight responses (i.e., habituate) to human approaches when the latter are benign [28]. On the other hand, more lateralized elks are bolder and more likely to move around, whereas less lateralized animals tend to remain near humans year-round [28].

Substantial individual variation in the strength of cerebral lateralization may be associated with individual variation in behaviour. For example, non-lateralized domestic chicks emitted more distress calls and took longer to resume pecking at food after exposure to a simulated predator than lateralized chicks [29]. Strongly lateralized convict cichlids (*Amatitlania nigrofasciata*) are quicker to emerge from a refuge indicative of boldness [30]. The degree of laterality seems to be positively correlated with stress reactivity in Port Jackson sharks (*Heterodontus portusjacksoni*) [31].

Recently, Whiteside and colleagues [32] showed that pheasants with a strong foot preference in motor tasks were more likely to die earlier in natural conditions than conspecifics with a mild foot preference. This study is the first trying to link lateralization with fitness in terms of survival and seems to suggest that the degree of lateralization does not linearly associate with benefits and that there is an optimum degree of laterality for pheasants in order to get the highest fitness (i.e., survival). Indeed, as stated by Rogers, Vallortigara and Andrew [1], arguing for computational advantages associated with the possession of an asymmetrical brain is not the same as arguing that the more asymmetric a brain, the more computationally-efficient it will be. In humans, there is a clear inverted U-shape curve in the relationship between degree of laterality and performance in word matching and face decision tasks [33], suggesting that a moderately asymmetrical brain would provide the greatest advantage. Finally, the relationship between lateralization and performance is task dependent [34]; therefore, a degree of laterality that may benefit one task may not benefit another. Survival requires an individual to detect predators, discriminate and handle food, cope with disease, navigate a complex environment, and learn strategies and much research on proxy measures of fitness looks at single factors, often in highly controlled environments.

3. Population-Level Lateralization as an Evolutionarily Stable Strategy (ESS)

Lateralization presents an intriguing aspect: it is often present at the population-level (i.e., directional asymmetry, where more than 50% of individuals within a population show the same direction of bias, such as handedness in humans, where about 90% of people are right-handed; [35]). If lateralization confers several advantages to the single individual in terms of brain efficiency, this cannot explain the alignment of the bias in the population.

The first evidence for a role of social behavior in population-level lateralization was provided by Rogers and Workman [36], who showed that more strongly lateralized chicks acquire a higher position in the social hierarchy than less lateralized chicks. Subsequently, Vallortigara and Rogers [7] reviewed the overall evidence and argued for a role of social interaction in the evolution of population-level brain asymmetry. The hypothesis was supported by a theoretical model developed by Ghirlanda and Vallortigara ([37]; see also [38]) showing that, in the context of prey-predator interactions, population–level lateralization can develop as an evolutionarily stable strategy (ESS) when individually asymmetrical organisms must coordinate their right-left behavioral patterns with those of other asymmetrical organisms. As a lateralized brain leads to behavioral biases when escaping from predators (e.g., [39]), the model considered the fitness consequences that the lateralization of one prey has when it interacts with other group-living prey subject to predation. The model assumed that the fitness was influenced by two contrasting selection pressures: (1) the benefit of being lateralized in the direction of the majority as a consequence of the "dilution effect" (i.e., prey in large groups have a lesser risk of being targeted by predators; [40]); and (2) the cost of being lateralized in the direction of the majority as a consequence of predators learning to anticipate prey escape strategies. In this second case, individuals who escape in a different direction from the majority have a benefit as they can surprise predators and survive more often. By varying the contribution of these costs and benefits, the model showed that population-level lateralization emerges as an ESS when neither of the two selection pressures is much stronger than the other. Thus, the successful strategy of group-living prey is to have a majority of individuals gaining protection from the group and escaping in the same direction when facing a predator and a minority of them being able to surprise the predator by escaping in the opposite direction. Empirical support for this hypothesis comes from fish schools, where animals showing the same turning bias as the majority of the group have an improved escape performance than fish at odds with the group [41].

A few years later, the mathematical model by Ghirlanda and Vallortigara [37] was extended by considering intraspecific interactions instead of interspecific prey-predator interactions [42]. Specifically, the new model considered the selective pressures of synergistic (cooperative) and antagonistic (competitive) interactions on individuals being lateralized in the same or opposite direction within the same species. It assumed that individuals lateralized in the same direction have a benefit in engaging in synergistic interactions as they can, for example, efficiently use the same tools or coordinate better. On the other side, individuals lateralized in the direction different from that of the majority have an advantage when engaging in antagonistic interactions for the same reason as in the previous model: they can surprise the opponent by adopting a strategy to which opponents are less accustomed. Empirical support for this assumption comes from the success of left-handers (i.e., lateralized in the opposite direction compared to the majority) in competitive sports such as fencing, boxing, and tennis (e.g., [43]; see also [44]). The ESS model for intraspecific interactions [42] showed that when the pressure of synergistic interactions becomes more and more important compared to that of antagonistic interactions, individually asymmetric organisms must interact with conspecifics and coordinate their activities and, consequently, asymmetry aligns in the majority of individuals in a population (i.e., directional or population-level asymmetry).

In order to provide empirical evidence for this prediction, the relationship between the level of lateralization and the presence of social behaviors was investigated using different species of bees as a model system (summarized in [45]; see also [46]). A series of experiments provided striking evidence that the alignment of lateralization within the population may be a consequence of social interactions frequently encountered during the course of evolution [47–49]. In fact, eusocial honeybees *Apis mellifera* [47], three species of primitively social Australian stingless bees [48], and annual social bumblebees *Bombus terrestris* [49], but not the solitary bees *Osmia rufa* [47], were found to be asymmetrical at the population-level for the use of the left and right antennae in recalling olfactory memories. However, all these studies investigated the use of a preferred antenna in recalling a learnt memory of an association between an odor and a food reward, and not really social interactions.

The first evidence of the role of antennal asymmetries in social interactions was shown in highly social ants *Formica rufa* [50]. By looking at "feeding" contacts where a "donor" ant exchanges food with a "receiver" ant through trophallaxis, the researchers [50] observed population-level asymmetry, with the "receiver" ant using the right antenna more frequently than the left antenna. The role of antennal asymmetries has also been investigated by observing the behavior of different dyads of honeybees with only the left, only the right, or both antennae in use, and belonging to the same or different hives [51]. In bees belonging to the same hive, dyads having only the right antenna in use took less time to get in contact and interacted more positively then dyads with only the left antennae, which instead interacted more aggressively than the other two groups. Interestingly, for bees belonging to different hives, dyads with only the right antenna in use displayed more aggressive interactions than bees with only the left or both antennae [51]. This suggests that the right antenna seems to control the correct behavioral response, depending on the social context, i.e., positive interactions between individuals of the same colony and negative interactions between individuals belonging to different colonies. A similar pattern of behavior between individuals of the same colony has been found in primitively social stingless bees *Trigona carbonaria*, where the right antenna stimulates positive contact and the left stimulates avoidance or attack [52].

Advantages of the population-level lateral bias have also been documented in the preference for keeping the mother on the left side in several terrestrial and aquatic mammal infants, supporting the idea of the role that lateralization plays in social interaction [53].

4. Individual- or Population-Level Lateralization as an ESS

Only recently, however, our research provided surprising findings: not only social species, but also so-called "non-social" species, of insects show asymmetries at the population-level when their limited interactions with others individuals are considered. This is the case for *Osmia rufa*, a species that does not show behavioral asymmetry in the recall of short-term olfactory memory [47], but shows population-level lateralization in aggressive displays [54], similarly to eusocial honeybees [51] and social stingless bees *T. carbonaria* [52]. Clearly, being engaged in interactions with other individuals, rather than the way in which the species nests (socially or not), may affect lateralization.

In honeybees, so far, all the identified biases occur at the population-level: in the use of the right visual pathway to learn visual stimuli [55], in the different use of the antennae in learning and recall of olfactory memories [56–58], and in context-dependent social interactions with conspecifics [51]. A recent study, however, suggests that honeybees tested in a tunnel with gaps of different apertures to the right and left sides, do not show population-level lateralization [59]. In this task, some individuals showed a bias to the right, some others a bias to the left, and yet others no bias. This may indicate that behavioral biases in bees vary in strength and direction, depending on whether the task requires coordination among individuals. Note, however, that very few bees showed individual bias in this task, and thus it is not clear whether individual lateralization was observed. Another example may be provided by foragers of *F. pratensis* ants, a species which does not use trail pheromones, moves more often to the left side than to the right whilst walking towards the nest, and does not show any bias when leaving the nest [60]. Moreover, it is still not clear to what extent the alignment occurs. Red wood ants *Formica rufa* belonging to different colonies show population-level biases in different directions when tested for forelimb preference during a gap crossing task, suggesting that social pressures act to coordinate individuals within the same colony and not necessarily at the species-level [61].

The different types of social interaction can generate evolutionarily stable strategies of lateralization at the individual- or population-level, depending on ecological contexts. Indeed, as we showed, population-level asymmetries have been observed in aggressive and mating displays in so-called "solitary" insects (e.g., tephrid flies, [62]; mason bees, [54]), suggesting that engagement in specific inter-individual interactions rather than "sociality" in general may generate population-level lateralization. This implies that lateralization is not necessarily a static feature of the neural organization, but is modulated by the functional context. For example, the nematode *C. elegans* exhibits a pronounced motor bias: males show a right-turning population-bias during mating. Interestingly, this motor bias is also observed in nematodes with mirror–reversed anatomical asymmetry, perhaps driven by epigenetic factors rather than by genetic variation [63].

The hypothesis that lateralization arises as an ESS is general and thus can predict either population- or individual-level lateralization, depending on the type of interactive behavior considered (e.g., cooperative or competitive) and ecological context. Although the advantage of being aligned in the same direction is clear in cooperative behavior, it is not in aggressive interactions. Indeed, it may be more advantageous for an aggressive display to not be directional, since population-level bias would also mean predictability [7]. For example, if an individual attacks another individual, it would be more convenient for it to be unpredictable. As a consequence, although each individual would have an (individual-level) bias, there will be 50:50 right:left-biased individuals in the population. This is the case for some predators, such as sailfish, which are lateralized at the individual-level in attacking schooling sardines on one side (and the stronger they are lateralized, the more successful they are at capturing their prey), but that overall, do not show a population-level bias [64]. However, if we think specifically about aggressive displays (and not the interactions), the alignment within the population may be linked to the need of an individual to position itself in a congruent way from a postural/motor point of view, as happens in mating (for a review, see [65]).

If being aligned in the same direction may help individuals to better coordinate with each other in specific tasks that require coordination between two or more individuals, being more or less biased in opposite directions may also have a potential benefit in other tasks where it is important to make best use of the available resources. This is something that future studies should address.

5. Conclusions

The ESS remains the single most powerful and widespread evolutionary hypothesis to explain lateralization. The ESS theoretical models [37,42] are well-supported by the new data showing population-level lateralization in interactions in the so-called "solitary" insects [54] and individual-level lateralization in social insects for tasks not requiring coordination [59], as the models predict that when social pressures become higher, population-level lateralization arises.

It is important, however, in order to avoid misunderstanding of the theory, to distinguish the claims of the ESS theory as an evolutionary hypothesis (i.e., in terms of natural history) and the claims concerning current living organisms. In terms of natural history, the ESS model hypothesizes that individual-level lateralization emerged first (because of the computational advantages associated with the individual possession of a slight asymmetry between the two halves of the brain) and that an alignment in the direction of the asymmetries evolved subsequently as a result of the interactions between individually-asymmetric organisms. For current-living organisms, any neat distinction between social and non-social species is obviously meaningless because definitions attain to the formal convention of specific disciplines. In entomology, honeybees are social and mason bees are solitary. But of course, this does not mean that mason bees do not interact with conspecifics. Thus, for the

Symmetry **2018**, *10*, 739

EES theory, the crucial issue is not the abstract definition of a species as social or not social, but rather whether a specific lateralized behavior entails constraints associated with the presence of other individuals performing the same lateralized behavior.

Importantly, the theory does not predict in a simplistic way that all living species that are not "social" should be lateralized at the individual level and that those that are "social" should be lateralized at the population level. What is important is the presence of inter-individual interactions in which the asymmetry of the individuals influences that of others (e.g., aggressive interactions in mason bees, [54]). In other words, a major prediction of ESS theory is that alignment in lateralization should be expected whenever asymmetric individuals exhibit a benefit from coordination with other asymmetric individuals which is higher than the cost associated with the predictability of their individual behavior. Vice versa, the lack of alignment in lateralization should be expected whenever costs associated with the predictability of individual behavior overcome the benefit of coordinating the behavior among different asymmetric individuals. This is a testable hypothesis that holds true, irrespective of whether individuals of a species are conventionally defined as "social" or "solitary".

For the ESS theory, individual-level and population-level lateralization are the two sides of the same coin or, even better, of the same continuum: the stability (i.e., an ESS) can be obtained with an individual-level or population-level asymmetry, depending on the context. In other words, the theory does not predict that social species need to be lateralized at the population-level, but rather that individual-level or population-level lateralization emerges as an ESS. Indeed, as discussed above, there may be cases in which, within the group, it is stable to be asymmetrical at the individual level (e.g., [64]).

Author Contributions: E.F. and G.V. conceived the paper, E.F. wrote the paper with inputs and additions from G.V.

Funding: This research received no external funding.

Acknowledgments: We thank Lesley Rogers for inviting us to contribute this paper in this special issue of Symmetry.

Conflicts of Interest: The authors declare no conflicts of interest.

References

1. Rogers, L.J.; Vallortigara, G.; Andrew, R. *Divided Brains: The Biology and Behaviour of Brain Asymmetries*, 1st ed.; Cambridge UP: Cambridge, UK, 2013.
2. Versace, E.; Vallortigara, G. Forelimb preferences in human beings and other species: Multiple models for testing hypotheses on lateralization. *Front. Psychol.* **2015**, *6*, 233. [CrossRef]
3. Rogers, L.J.; Vallortigara, G. When and Why Did Brains Break Symmetry? *Symmetry* **2015**, *7*, 2181–2194. [CrossRef]
4. Vallortigara, G.; Versace, E. Laterality at the Neural, Cognitive, and Behavioral Levels. In *APA Handbook of Comparative Psychology: Vol. 1. Basic Concepts, Methods, Neural Substrate, and Behavior*; American Psychological Association: Washington, DC, USA, 2017; pp. 557–577.
5. Frasnelli, E.; Vallortigara, G.; Rogers, L.J. Left-right asymmetries of behavioural and nervous system in invertebrates. *Neurosci. Biobehav. Rev.* **2012**, *36*, 1273–1291. [CrossRef]
6. Frasnelli, E. Brain and behavioral lateralization in invertebrates. *Front. Psychol.* **2013**, *4*, 1–10. [CrossRef]
7. Vallortigara, G.; Rogers, L.J. Survival with an asymmetrical brain: Advantages and disadvantages of cerebral lateralization. *Behav. Brain Sci.* **2005**, *28*, 575–588. [CrossRef]
8. Vallortigara, G.; Rogers, L.J.; Bisazza, A. Possible evolutionary origins of cognitive brain lateralization. *Brain Res. Rev.* **1999**, *30*, 164–175. [CrossRef]
9. Vallortigara, G. Comparative neuropsychology of the dual brain: A stroll through left and right animals' perceptual worlds. *Brain Lang.* **2000**, *73*, 189–219. [CrossRef]
10. Andrew, R.J. (Ed.) *Neural and Behavioural Plasticity: The Use of the Domestic Chicken as a Model*; Oxford University Press: Oxford, UK, 1991.

11. Vallortigara, G.; Cozzutti, C.; Tommasi, L.; Rogers, L.J. How birds use their eyes: Opposite left-right specialisation for the lateral and frontal visual hemifield in the domestic chick. *Curr. Biol.* **2001**, *11*, 29–33. [CrossRef]

12. Rogers, L.J.; Anson, J.M. Lateralisation of function in the chicken forebrain. *Pharmacol. Biochem. Behav.* **1979**, *10*, 679–686. [CrossRef]

13. Valenti, A.; Sovrano, V.A.; Zucca, P.; Vallortigara, G. Visual lateralization in quails. *Laterality* **2003**, *8*, 67–78. [CrossRef]

14. Rogers, L.J.; Zucca, P.; Vallortigara, G. Advantage of having a lateralized brain. *Proc. R. Soc. Lond. B* **2004**, *271*, S420–S422. [CrossRef]

15. McGrew, W.C.; Marchant, L.F. Laterality of hand use pays off in foraging success for wild chimpanzees. *Primates* **1999**, *40*, 509–513. [CrossRef]

16. Marcinowski, E.C.; Campbell, J.M.; Faldowski, R.A.; Michel, G.F. Do hand preferences predict stacking skill during infancy? *Dev. Psychobiol.* **2016**, *58*, 958–967. [CrossRef]

17. Magat, M.; Brown, C. Laterality enhances cognition in Australian parrots. *Proc. R. Soc. Lond. B* **2009**, *276*, 4155–4162. [CrossRef]

18. Rogers, L.J. Evolution of hemispheric specialisation: Advantages and disadvantages. *Brain Lang.* **2000**, *73*, 236–253. [CrossRef]

19. Güntürkün, O.; Diekamp, B.; Manns, M. Asymmetry pays: Visual lateralization improves discrimination success in pigeons. *Curr. Biol.* **2000**, *10*, 1079–1081. [CrossRef]

20. Bell, A.T.; Niven, J.E. Individual-level, context-dependent handedness in the desert locust. *Curr. Biol.* **2014**, *24*, R382–R383. [CrossRef]

21. Bell, A.T.; Niven, J.E. Strength of forelimb lateralization predicts motor errors in an insect. *Biol. Lett.* **2016**, *12*, 20160547. [CrossRef]

22. Miler, K.; Kuszewska, K.; Woyciechowski, M. Larval antlions with more pronounced behavioural asymmetry show enhanced cognitive skills. *Biol. Lett.* **2017**, *13*, 20160786. [CrossRef]

23. Hook, M.A.; Rogers, L.J. Visuospatial reaching preferences of common marmosets (*Callithrix jacchus*): An assessment of individual biases across a variety of tasks. *J. Comp. Psychol.* **2008**, *122*, 41–51. [CrossRef]

24. Brown, C.; Braithwaite, V.A. Effects of predation pressure on the cognitive ability of the poeciliid *Brachyraphis episcopi*. *Behav. Ecol.* **2005**, *16*, 482–487. [CrossRef]

25. Krakauer, A.H.; Blundell, M.A.; Scanlan, T.N.; Wechsler, M.S.; McCloskey, E.A.; Yu, J.H.; Patricelli, G.L. Successfully mating male sage-grouse show greater laterality in courtship and aggressive interactions. *Anim. Behav.* **2016**, *111*, 261–267. [CrossRef]

26. Jennings, D.J. Right-sided bias in fallow deer terminating parallel walks: Evidence for lateralization during a lateral display. *Anim. Behav.* **2012**, *83*, 1427–1432. [CrossRef]

27. Camerlink, I.; Menneson, S.; Turner, S.P.; Farish, M.; Arnott, G. Lateralization influences contest behaviour in domestic pigs. *Sci. Rep.* **2018**, *8*, 12116. [CrossRef]

28. Found, R.; Clair, C.C. Ambidextrous ungulates have more flexible behaviour, bolder personalities and migrate less. *R. Soc. Opensci.* **2017**, *4*, 160958. [CrossRef]

29. Dharmaretnam, M.; Rogers, L.J. Hemispheric specialization and dual processing in strongly versus weakly lateralized chicks. *Behav. Brain Res.* **2005**, *162*, 62–70. [CrossRef]

30. Reddon, A.R.; Hurd, P.L. Individual differences in cerebral lateralization are associated with shy-bold variation in the convict cichlid. *Anim. Behav.* **2009**, *77*, 189–193. [CrossRef]

31. Byrnes, E.E.; Pouca, C.V.; Brown, C. Laterality strength is linked to stress reactivity in Port Jackson sharks (*Heterodontus portusjacksoni*). *Behav. Brain Res.* **2016**, *305*, 239–246. [CrossRef]

32. Whiteside, M.A.; Bess, M.M.; Frasnelli, E.; Beardsworth, C.E.; Langley, E.J.G.; van Horik, J.O.; Madden, J.R. Low survival of strongly footed pheasants may explain constraints on lateralization. *Sci. Rep.* **2018**, *8*, 13791. [CrossRef]

33. Hirnstein, M.; Leask, S.; Rose, J.; Hausmann, M. Disentangling the relationship between hemispheric asymmetry and cognitive performance. *Brain Cogn.* **2010**, *73*, 119–127. [CrossRef]

34. Boles, D.B.; Barth, J.M.; Merrill, E.C. Asymmetry and performance: Toward a neurodevelopmental theory. *Brain Cogn.* **2008**, *66*, 124–139. [CrossRef]

35. McManus, I.C. *Right Hand, Left Hand: The Origins of Asymmetry in Brains, Bodies, Atoms, and Cultures*; Weidenfeld & Nicolson: London, UK, 2002.

36. Rogers, L.J.; Workman, L. Light exposure during incubation affects competitive behaviour in domestic chicks. *Appl. Anim. Behav. Sci.* **1989**, *23*, 187–198. [CrossRef]
37. Ghirlanda, S.; Vallortigara, G. The evolution of brain lateralisation: A game-theoretical analysis of population structure. *Proc. R. Soc. B* **2004**, *271*, 853–857. [CrossRef]
38. Vallortigara, G. The evolutionary psychology of left and right: Costs and benefits of lateralization. *Dev. Psychobiol.* **2006**, *48*, 418–427. [CrossRef]
39. Lippolis, G.; Bisazza, A.; Rogers, L.J.; Vallortigara, G. Lateralization of predator avoidance responses in three species of toads. *Laterality* **2002**, *7*, 163–183. [CrossRef]
40. Foster, W.A.; Treherne, J.E. Evidence for the dilution effect in the selfish herd from fish predation of a marine insect. *Nature* **1981**, *293*, 508–510. [CrossRef]
41. Chivers, D.P.; McCormick, M.I.; Allan, B.J.; Mitchell, M.D.; Goncalves, E.J.; Bryshun, R.; Ferrari, M.C. At odds with the group: Changes in lateralization and escape performance reveal conformity and conflict in fish schools. *Proc. R. Soc. Lond. B* **2016**, *283*, 20161127. [CrossRef]
42. Ghirlanda, S.; Frasnelli, E.; Vallortigara, G. Intraspecific competition and coordination in the evolution of lateralization. *Phil. Trans. R. Soc. Lond. B* **2009**, *364*, 861–866. [CrossRef]
43. Loffing, F. Left- handedness and time pressure in elite interactive ball games. *Biol. Lett.* **2017**, *13*, 20170446. [CrossRef]
44. Faurie, C.; Raymond, M. Handedness, homicide and negative frequency-dependent selection. *Proc. R. Soc. Lond. B* **2005**, *272*, 25–28. [CrossRef]
45. Frasnelli, E.; Haase, A.; Rigosi, E.; Anfora, G.; Rogers, L.J.; Vallortigara, G. The bee as a model to investigate brain and behavioural asymmetries. *Insects* **2014**, *5*, 120–138. [CrossRef]
46. Niven, J.E.; Frasnelli, E. Insights into the evolution of lateralization from the insects. *Progr. Brain Res.* **2018**, *238*, 3–31.
47. Anfora, G.; Frasnelli, E.; Maccagnani, B.; Rogers, L.J.; Vallortigara, G. Behavioural and electrophysiological lateralization in a social (*Apis mellifera*) but not in a non-social (*Osmia cornuta*) species of bee. *Behav. Brain Res.* **2010**, *206*, 236–239. [CrossRef]
48. Frasnelli, E.; Vallortigara, G.; Rogers, L.J. Right-left antennal asymmetry of odour memory recall in three species of Australian stingless bees. *Behav. Brain Res.* **2011**, *224*, 121–127. [CrossRef]
49. Anfora, G.; Rigosi, E.; Frasnelli, E.; Trona, F.; Vallortigara, G. Lateralization in the invertebrate brain: Left–right asymmetry of olfaction in bumble bee, *Bombus terrestris*. *PLoS ONE* **2011**, *6*, e18903. [CrossRef]
50. Frasnelli, E.; Iakovlev, I.; Reznikova, Z. Asymmetry in antennal contacts during trophallaxis in ants. *Behav. Brain Res.* **2012**, *232*, 7–12. [CrossRef]
51. Rogers, L.J.; Rigosi, E.; Frasnelli, E.; Vallortigara, G. A right antenna for social behaviour in honeybees. *Sci. Rep.* **2013**, *3*, 2045. [CrossRef]
52. Rogers, L.J.; Frasnelli, E. Antennal Asymmetry in Social Behavior of the Australian Stingless Bee, *Tetragonula carbonaria*. *J. Insect Behav.* **2016**, *29*, 491–499. [CrossRef]
53. Karenina, K.; Giljov, A.; Ingram, J.; Rowntree, V.J.; Malashichev, Y. Lateralization of mother–infant interactions in a diverse range of mammal species. *Nat. Ecol. Evol.* **2017**, *1*, 0030. [CrossRef]
54. Rogers, L.J.; Frasnelli, E.; Versace, E. Lateralized antennal control of aggression and sex differences in red mason bees, *Osmia bicornis*. *Sci. Rep.* **2016**, *6*, 29411. [CrossRef]
55. Letzkus, P.; Boeddeker, N.; Wood, J.T.; Zhang, S.W.; Srinivasan, M.V. Lateralization of visual learning in the honeybee. *Biol. Lett.* **2008**, *4*, 16–19. [CrossRef]
56. Letzkus, P.; Ribi, W.A.; Wood, J.T.; Zhu, H.; Zhang, S.W.; Srinivasan, M.V. Lateralization of olfaction in the honeybee *Apis mellifera*. *Curr. Biol.* **2006**, *16*, 1471–1476. [CrossRef]
57. Rogers, L.J.; Vallortigara, G. From antenna to antenna: Lateral shift of olfactory memory recall by honeybees. *PLoS ONE* **2008**, *3*, e2340. [CrossRef]
58. Frasnelli, E.; Vallortigara, G.; Rogers, L.J. Response competition associated with right-left antennal asymmetries of new and old olfactory memory traces in honeybees. *Behav. Brain Res.* **2010**, *209*, 36–41. [CrossRef]
59. Ong, M.; Bulmer, M.; Groening, J.; Srinivasan, M.V. Obstacle traversal and route choice in flying honeybees: Evidence for individual handedness. *PLoS ONE* **2017**, *12*, e0184343. [CrossRef]
60. Hönicke, C.; Bliss, P.; Moritz, R.F. Effect of density on traffic and velocity on trunk trails of *Formica pratensis*. *Sci. Nat.* **2015**, *102*, 17. [CrossRef]

61. Calcraft, P.R.T.; Bell, A.T.; Husbands, P.; Philippides, A.; Niven, J.E. The evolution of handedness: Why are ant colonies left-and right-handed? *Biomath Commun.* **2016**, *3*. [CrossRef]
62. Benelli, G.; Donati, E.; Romano, D.; Stefanini, C.; Messing, R.H.; Canale, A. Lateralisation of aggressive displays in a tephritid fly. *Sci. Nat.* **2015**, *102*, 1. [CrossRef]
63. Downes, J.C.; Birsoy, B.; Chipman, K.C.; Rothman, J.H. Handedness of a motor program in *C. elegans* is independent of left-right body asymmetry. *PLoS ONE* **2012**, *7*, e52138. [CrossRef]
64. Kurvers, R.H.; Krause, S.; Viblanc, P.E.; Herbert-Read, J.E.; Zaslansky, P.; Domenici, P.; Couillaud, P. The evolution of lateralization in group hunting sailfish. *Curr. Biol.* **2017**, *27*, 521–526. [CrossRef]
65. Frasnelli, E. Lateralization in Invertebrates. In *Lateralized Brain Functions: Methods in Human and Non-Human Species*; Neuromethods; Springer Protocols; Rogers, L.J., Vallortigara, G., Eds.; Humana Press: New York, NY, USA, 2017; Volume 122, pp. 153–208.

symmetry

MDPI

Article

Social Risk Dissociates Social Network Structure across Lateralized Behaviors in Spider Monkeys

Emily R. Boeving * and Eliza L. Nelson

Department of Psychology, Florida International University, Miami, FL, 33199, USA; elnelson@fiu.edu
* Correspondence: eboev001@fiu.edu; Tel.: +1-305-348-4032

Received: 1 August 2018; Accepted: 6 September 2018; Published: 9 September 2018

Abstract: Reports of lateralized behavior are widespread, although the majority of findings have focused on the visual or motor domains. Less is known about laterality with regards to the social domain. We previously observed a left-side bias in two social affiliative behaviors—embrace and face-embrace—in captive Colombian spider monkeys (*Ateles fusciceps rufiventris*). Here we applied social network analysis to laterality for the first time. Our findings suggest that laterality influences social structure in spider monkeys with structural differences between networks based on direction of behavioral bias and social interaction type. We attribute these network differences to a graded spectrum of social risk comprised of three dimensions.

Keywords: social networks; laterality; social behavior; spider monkey; risk; social interaction

1. Introduction

Reports of lateralized behavior are widespread, particularly in the visual and motor domains [1,2]. Decades of research have led to the general consensus that behavioral lateralization is subserved by asymmetric brain function. These brain-behavior asymmetries may serve to streamline neurobiological processes, thereby increasing behavioral efficiency in unpredictable or arousing situations, such as social interactions [3,4]. Thus, laterality may be particularly advantageous in gregarious species such as primates.

In a recent synthesis of prior research, Rogers and Vallortigara [1] linked left biases in social behavior to the right hemisphere as a general pattern of lateralization in vertebrates. However, we later showed that not all social behaviors are associated with this pattern of laterality [5]. Specifically, we found that two variations of embracing, but not grooming, were lateralized in Colombian spider monkeys. We argued that the differences in lateralization in social affiliative behaviors were due to the social dynamic in which these behaviors occurred, with grooming considered a low-stakes routine state while embraces were high-stakes risky events. In this study, we focused on assessing the behavioral patterns among individuals within a group, and did not take into account the relational patterns of the group as a whole (e.g., interaction history). While consistent with other laterality investigators, this reductionist approach does not capture the true dynamics of a social system, begging the question: does laterality influence social structure?

Spider monkeys are one of a handful of primates living in fission-fusion [6], a social dynamic defined by separations and reunions. Embraces are a contact greeting gesture that occur at the time of reunions in spider monkeys [7]. In the standard embrace, the hands are wrapped around the body and the face is placed along the trunk [7,8]. A variation is the face-embrace, in which faces touch [5]. Fission-fusion is characterized by marked unpredictability and low social cohesion compared with species that have a known stable hierarchy, cohesive social groups, and low variability in interactive exchanges [9,10]. With these differences in mind, social interactions within species living in fission-fusion may consist of a level of risk unlike that experienced in other social dynamics, and laterality may play a role in negotiating this risk [2]. In general, social behavior in fission-fusion species is remarkably multi-dimensional, and can be difficult to tease apart.

One method for teasing apart complex social systems is social network analysis [11], a concept with roots in the mathematical field of graph theory. Social network analysis is a tool used to compute and visualize structural relationships in relational data. There is a long history of applying network analysis in the study of sociality in primates (for a review, see [12]) and other species [13]. Yet social network analysis has never been applied in the area of behavioral laterality. Network analysis alone has the unique ability to characterize and mathematically represent global inter-connected elements [14]. Within behavioral laterality, network level information may provide a more sophisticated method to examine topological patterns that represent potential advantages of laterality for behavior, and to accurately depict the multi-dimensional nature of social interaction.

As our primary objective, we leveraged social network analysis in the dataset reported by Boeving, Belnap and Nelson [5] to examine whether similarly lateralized behaviors (i.e., embrace and face-embrace) also have similar network structures, and we predicted that these networks would not differ. In our secondary objective, we examined social networks based on direction of laterality (i.e., left or right) regardless of behavior type by pooling embrace and face-embrace into an affiliative category. We hypothesized that laterality would influence network structure, and we predicted that global left and right affiliative networks would diverge. Finally, we examined the influence of both direction of laterality and behavior type on social network structure by creating four sub-networks of left embrace, left face-embrace, right embrace, and right face-embrace. We hypothesized that laterality, but not behavior type, would alter network structure. We predicted that the left sub-networks would differ from the right sub-networks, but that sub-networks within a behavior (i.e., embrace or face-embrace) would not differ.

2. Materials and Methods

2.1. Social Network Construction from Live Coded Behavior

We constructed social networks from live coded behavioral observations of 15 captive Colombian spider monkeys (*Ateles fusciceps rufiventris*). Portions of these data were previously reported in Boeving, Belnap and Nelson [5]. To briefly summarize, 186 h of data were captured between May and August 2015 using the Animal Behaviour Pro mobile iOS application on apple iPod 5th generation [15]. The application was programmed with information about the individual monkeys to capture initiators and receivers of embrace and face-embrace with the modifier set as side (i.e., left or right positioning). Left or right was recorded with reference to the positioning of the faces regardless of whether there was contact or not. Directionality was not determined by any positioning of the limbs. Data were collected using the continuous sampling method, and *ad libitum* recording method [16,17] so that all occurrences of the target behaviors could be captured across three equally distributed time periods throughout the day to avoid disruptions due to husbandry procedures. The DuMond Conservancy Institutional Animal Care and Use Committee approved the research, and the study was conducted in accordance with the laws of the United States. The research adhered to the American Society of Primatologists (ASP) Principles for the Ethical Treatment of Non-Human Primates.

2.2. Social Network Analysis

We utilized social network analysis as the computational method to investigate potential structural differences within all networks. Networks were computed and visualized in Cytoscape (http://www.cytoscape.com) (Version 3.4.0; [18]), an open source software project for modeling interaction networks. The network metric of *degree centrality*, which provides a composite score from the *in-degree* value (i.e., interactions directed towards a monkey) and *out-degree* value (i.e., interactions directed by a monkey to others), was examined because this metric quantifies the number of edges (i.e., social interactions) shared between nodes (i.e., monkeys). The degree centrality of node (v) for a given graph (G) = ($\mathcal{V}$, E) with $|\mathcal{V}|$ nodes and $|E|$ edges defined as:

$$C_D (v) = deg (v)$$

Using the metric degree centrality, the total number of interactions for each individual was computed where monkeys with the most connected interactions (initiated or received) were positioned in the center of the graph and monkeys with fewer connected interactions were positioned along the perimeter. Within Cytoscape, we used a variant of the "Kamada-Kawai Algorithm," a spring-embedded algorithm that forces connected nodes together while also forcing disconnected nodes away from the center [19]. We constructed weighted networks because this method is best suited for graphically representing the variation in social bonds [20,21]. All edges were weighted based on frequency of interaction with thicker edges denoting more interactions and thinner edges denoting fewer interactions. Node size denotes variation in rank of degree centrality where larger nodes indicate higher values of degree centrality and smaller nodes indicate lower values of degree centrality.

2.3. Statistical Analysis

To examine whether similarly lateralized behaviors (i.e., embrace and face-embrace) have similar network structures, we first pooled frequency data from each behavior separately regardless of side to create global embrace and global face-embrace networks. To investigate the potential effect of laterality on social network structure, we then pooled affiliative frequency data according to side of positioning to create global left affiliative and global right affiliative networks. Finally, we examined the effect of laterality within each type of embrace by constructing four direction x behavior networks: left embrace, right embrace, left face-embrace, and right face-embrace. *t*-Tests and ANOVA with post hoc comparisons were used to compare the resulting networks.

3. Results

A total of 1623 social interactions were examined. Of these, 1270 were embraces and 353 were face-embraces, corresponding to 1227 left affiliative and 396 right affiliative interactions. Individual raw frequency scores for each behavior are reported in Table A1. Four juveniles were excluded from further analysis due to multiple zero values for out-degree, which we suggest is age-related and would not accurately portray degree centrality in the spider monkey group. Network degree centrality values for the global comparisons can be found in Table 1. Unpaired *t*-tests found a significant difference in degree centrality between the global embrace and face-embrace networks ($t(28) = 3.43$, $p < 0.01$, $d = 1.296$; Figure A1), and a significant difference in degree centrality between the global left and right affiliative networks ($t(20) = 3.92$, $p < 0.001$, $d = 1.753$; Figure A2). There was no sex difference in the global left affiliative, global right affiliative, or global embrace networks (all $p > 0.05$). However, there was a sex difference in the face-embrace network such that females initiated the face-embrace behavior more than males, and males received more of these interactions compared to females ($F(1,13) = 4.82$, $p < 0.05$, $\eta^2 = 0.270$). To further examine structural differences between embrace and face-embrace within the context of laterality, we examined the four sub-networks (left embrace, right embrace, left face-embrace, right face-embrace). ANOVA revealed a significant difference in degree centrality among the sub-networks ($F(3,40) = 20.72$, $p < 0.001$, $\eta^2 = 0.608$; Figure 1). Post hoc analyses found that each sub-network was different from the others (all $p < 0.05$).

Table 1. Individual degree centrality values.

Monkey	Sex	Left Affiliative	Right Affiliative	Embrace	Face-Embrace
Bon Jovi (Bon)	M	202	57	214	62
Butch (Bu)	M	294	82	263	128
Carmelita (Carm)	F	76	25	82	24
Cleo	F	208	62	208	73
CJ	F	108	32	123	19
Dusky (Dusk)	F	164	46	191	31
Mason (Mas)	M	372	104	342	141
Mints (Min)	F	79	38	136	4
Molly (Mol)	F	94	25	110	15
Sunday (Sun)	M	261	101	296	83
Uva	M	386	144	445	121

M = Male, F = Female. The higher the degree centrality value, the more highly connected a monkey is to others.

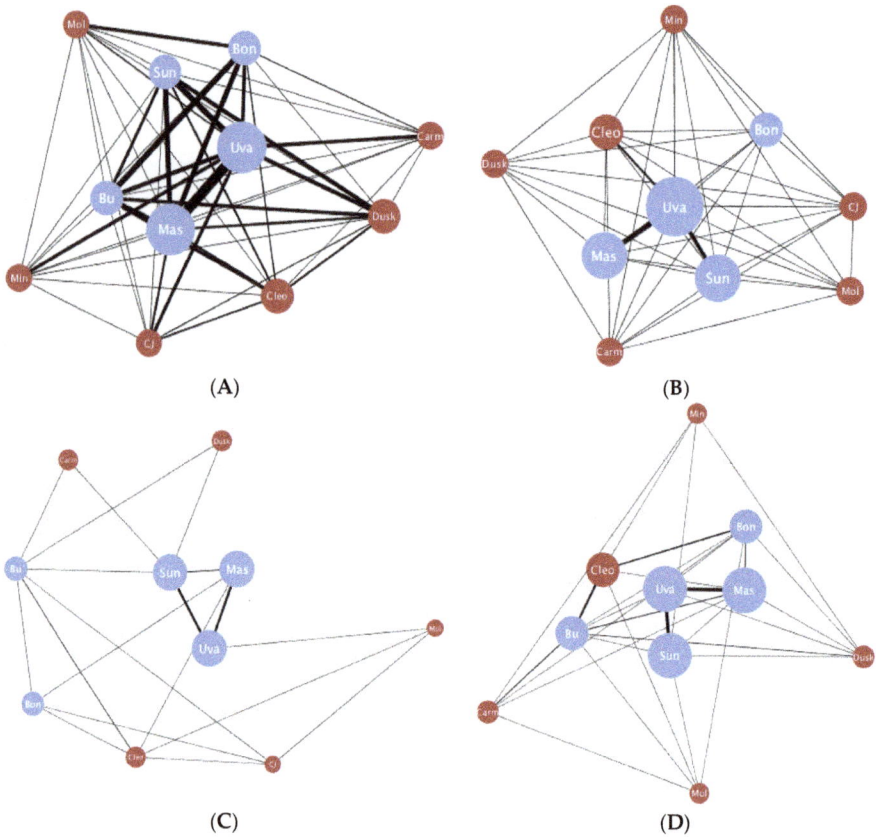

(A)

(B)

(C)

(D)

Figure 1. Clockwise from top left: (**A**) Left embrace; (**B**) Right embrace; (**C**) Left face-embrace; and (**D**) Right face-embrace. Networks are ordered on social risk index (see text for details). Red denotes females, and blue denotes males. Nodes are weighted such that the larger the node, the higher the degree centrality. Edges are weighted such that thickness denotes frequency of interactions.

4. Discussion

The primary objective of this study was to examine if behaviors with similar patterns of behavioral laterality would also have similar social network structures. We examined the social affiliative behaviors, embrace and face-embrace, which we previously have shown to be left lateralized in spider monkey behavior [5]. Contrary to our predictions, we found that the network for embrace was structurally different from that of face-embrace. We then explored our secondary objective examining whether the side with which the social affiliative behaviors were performed had an effect on network structure. Here our results confirmed our prediction that the global left affiliative network was structurally different from the global right affiliative network. Finally, our analysis of sub-networks parsing direction within each behavior partially supported our prediction. All four sub-networks were different from each other, suggesting an interaction between laterality and behavior type. We discuss these differences in social network structure in the context of three dimensions of social risk.

The concept of risk is often described in the non-human primate literature in the context of risk of aggression from neighboring groups [22], predation [23], and loss of resources [24], all of which are typical challenges for species living in the wild. Rebecchini et al. [25] first identified embracing as a component of risk in spider monkeys, and Boeving, Belnap and Nelson [5] suggested that embrace risk may be graded according to the type of physical contact with face-embrace having higher risk given the close placement of the faces. By comparison, embrace is lower risk because the faces do not touch. Here, we label this type of risk *contact risk*. Although embrace and face-embrace have a similar left behavioral lateralization pattern, the finding that they do not have similar network structures supports the conclusion that these behaviors are related but distinct. The graphical representation of the embrace network conveys the robustness of this behavior (Figure A1A). Specifically, most individuals engaged in embracing, and with high frequencies, yielding a network graph with most monkeys having high values for degree centrality. Overall, this pattern indicates strong cohesion in the embrace network. In contrast, the face-embrace network depicts interactive patterns in which only a few males were strongly bonded (Figure A1B). When in-degree and out-degree were examined, both males and females initiated and received within the embrace network, but there was a significant difference in the face-embrace network where females initiated more face-embrace and males received more of this behavior. This sex difference is notable because aggression towards females from male spider monkeys is a known pattern [26], making the social lives of female spider monkeys especially risky. In captivity, intra-group aggression is an important consideration given that wild female spider monkeys emigrate from their natal group [26,27]. We envisioned the face-embrace to be the riskier of the two embraces given the close face contact. Yet, with the known pattern of aggression towards females in mind, our social network analysis points to a second aspect of social risk within the face-embrace: *partner risk*. Social risk in relation to sex roles has been widely discussed in the human literature. For example, female sexual risk taking within certain communities is associated with greater risk of male aggression towards them [28,29]. Contact and partner variables have also been examined in the literature on social touch laterality in human kissing [30–34] and embracing [35,36], although these studies have not framed their findings in the context of risk, which may be an avenue in the future to connect these two streams of research.

A third type of risk identified by our network analyses is *laterality risk*. This dimension of risk was informed by our analyses that identified a structural difference between the global left affiliative and global right affiliative networks. In the left affiliative network, several monkeys were central. In contrast, the right affiliative network had a significantly different architecture in which fewer monkeys were central to the network, and in which the behavior occurred less frequently. Previous work has suggested that the right hemisphere plays an important role in the monitoring and detection of uncertain events in the environment, while the left hemisphere is more involved in routine behavior [2]. This role differentiation between hemispheres is particularly relevant when considering the positioning of the body for embrace and face-embrace. Specifically, if the functional split between hemispheres is correct, then positioning others on the right side for either behavior would be risky. Moreover, face-embrace would be especially risky given the close contact of the face coupled with the hypothesized decrease in ability for social monitoring when engaging others on the right side. It would thus be advantageous to position conspecifics on the left side given the hypothesized neural processing benefit. In line with this hypothesis, the structure of the left lateralized affiliative network pattern can be characterized as a highly cohesive network where all monkeys engaged in the behavior, and engaged frequently (Figure A2A). In contrast, the right lateralized network was lower in cohesion; engagement occurred less frequently, with only a few monkeys reaching high values of degree centrality (Figure A2B). Although not recorded in this study, capturing the sequence of behaviors that follow these risky interactions would further test this theory, and is a goal for future work.

Although we collected data over a four-month period, one limitation of this study is that we were not able to assess the stability of these networks over time. Juvenile data were excluded from analyses due to the low frequency of engagement in the behaviors we examined. However, we would expect this pattern to change as individuals mature and develop social bonds. The novel application of social network analysis could quantify this process, not only in primates, but other highly social species. Moreover, here we have utilized a between-networks approach based on our research question, but a within-networks approach across two or more timepoints could provide information about how an individual's position in a network changes as a function of development. A developmental network approach would also broaden our knowledge of the factors that contribute to the emergence of social laterality and its function.

Taken together, the structural differences between the four sub-networks confirmed a graded spectrum of social risk in spider monkeys along the three dimensions of risk: contact, partner, and laterality (Table 2). The sub-network with the lowest risk (i.e., left embrace) had the most participation and strongest cohesion, whereas the sub-network with the highest risk (i.e., right face-embrace) had the least participation and was the most disjointed of the networks indicating low cohesion (Figure 1). To answer our original question posed in the introduction, these findings suggest that laterality influences social structure. However, we acknowledge that social structure may also influence laterality, or that the relationship is bidirectional. Future work using longitudinal designs may address this point. Additional studies should also aim to include network analyses of other behavioral domains that could be related to laterality, such as cognition and motor skill. In conclusion, social network analysis is an exciting new avenue for characterizing brain-behavior relationships. In using this unique computational method to elucidate factors that drive global differences in social network topology, we advance our understanding of laterality within a social framework.

Table 2. Dimensions of social risk.

Behavior	Laterality	Contact	Partner	Risk Index
Left Embrace	Low	Low	Low	Lowest
Right Embrace	High	Low	Low	Mild
Left Face-Embrace	Low	High	High	Moderate
Right Face-Embrace	High	High	High	Highest

See text for details.

Author Contributions: Conceptualization, E.R.B. and E.L.N.; Analysis, E.R.B.; Writing—original draft, E.R.B.; Writing—review & editing, E.R.B. and E.L.N.

Funding: This research received no external funding.

Acknowledgments: We thank Monkey Jungle for supporting this project and members of the HANDS Lab for their assistance with data collection, and Starlie Belnap for her input on the statistical analysis. Alyssa Seidler provided the drawings in the graphical abstract. This is DuMond Conservancy publication no. 60.

Conflicts of Interest: The authors declare no conflict of interest.

Appendix

Table A1. Individual Raw Frequency Scores.

Monkey	Sex	Embrace	Face-Embrace
Bon Jovi (Bon)	M	92	16
Butch (Bu)	M	107	39
Carmelita (Carm)	F	36	17
Cleo	F	126	63
CJ	F	78	14
Dusky (Dusk)	F	92	27
Mason (Mas)	M	181	61
Mints (Min)	F	47	2
Molly (Mol)	F	81	11
Sunday (Sun)	M	151	22
Uva	M	198	80

M = Male, F = Female. Frequency is summed across interactions where the monkey initiated or received the behavior.

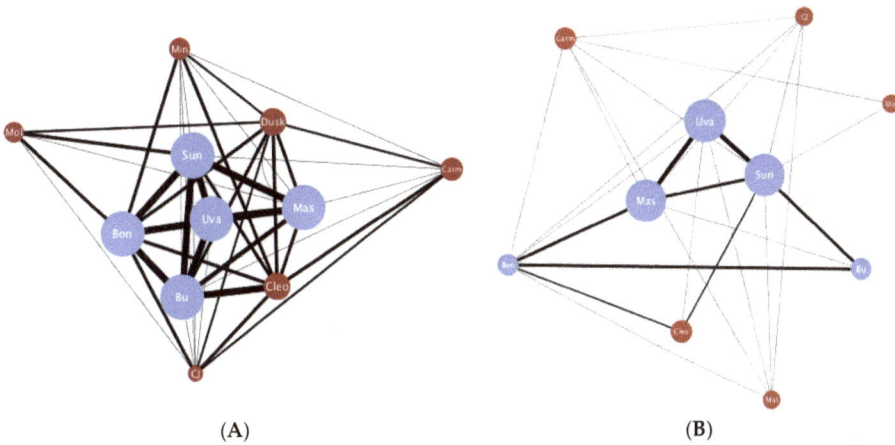

(A) (B)

Figure A1. Global embrace and global face-embrace networks differ.

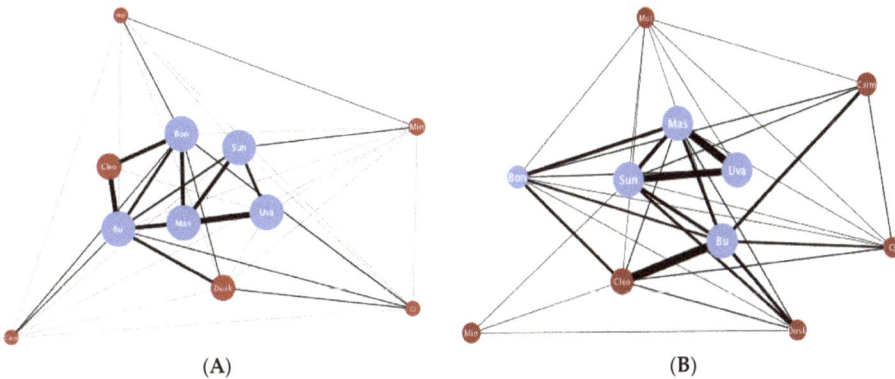

(A) (B)

Figure A2. Global left affiliative and global right affiliative networks differ. Red denotes females, and blue denotes males. Nodes are weighted such that the larger the node, the higher the degree centrality. Edges are weighted such that thickness denotes frequency of interactions.

References

1. Rogers, L.J.; Vallortigara, G. When and why did brains break symmetry? *Symmetry* **2015**, *7*, 2181–2194. [CrossRef]
2. MacNeilage, P.F.; Rogers, L.J.; Vallortigara, G. Origins of the left & right brain. *Sci. Am.* **2009**, *301*, 60–67. [PubMed]
3. Rogers, L.J.; Vallortigara, G.; Andrew, R.J. *Divided Brains: The Biology and Behaviour of Brain Asymmetries*; Cambridge University Press: Cambridge, UK, 2013.
4. Vallortigara, G.; Rogers, L.J. Survival with an asymmetrical brain: Advantages and disadvantages of cerebral lateralization. *Behav. Brain Sci.* **2005**, *28*, 575–588. [CrossRef] [PubMed]
5. Boeving, E.R.; Belnap, S.C.; Nelson, E.L. Embraces are lateralized in spider monkeys (*Ateles fusciceps rufiventris*). *Am. J. Primatol.* **2017**, *79*, e22654. [CrossRef] [PubMed]
6. Aguilar-Melo, A.R.; Calmé, S.; Smith-Aguilar, S.E.; Ramos-Fernandez, G. Fission-fusion dynamics as a temporally and spatially flexible behavioral strategy in spider monkeys. *Behav. Ecol. Sociobiol.* **2018**, *72*, 150. [CrossRef]
7. Schaffner, C.M.; Aureli, F. Embraces and grooming in captive spider monkeys. *Int. J. Primatol.* **2005**, *26*, 1093–1106. [CrossRef]
8. Eisenberg, J.F. Communication mechanisms and social integration in the black spider monkey, ateles fusciceps robustus, and related species. *Smithson. Contrib. Zool.* **1976**, *213*, 1–108. [CrossRef]
9. Aureli, F.; Schaffner, C.M.; Boesch, C.; Bearder, S.K.; Call, J.; Chapman, C.A.; Connor, R.; Di Fiore, A.; Dunbar, R.I.; Henzi, S.P. Fission-fusion dynamics. *Curr. Anthropol.* **2008**, *49*, 627–654. [CrossRef]
10. Ramos-Fernandez, G.; King, A.J.; Beehner, J.C.; Bergman, T.J.; Crofoot, M.C.; Di Fiore, A.; Lehmann, J.; Schaffner, C.M.; Snyder-Mackler, N.; Zuberbühler, K. Quantifying uncertainty due to fission–fusion dynamics as a component of social complexity. *Proc. R. Soc. B* **2018**, *285*, 20180532. [CrossRef] [PubMed]
11. Sueur, C.; Jacobs, A.; Amblard, F.; Petit, O.; King, A.J. How can social network analysis improve the study of primate behavior? *Am. J. Primatol.* **2011**, *73*, 703–719. [CrossRef] [PubMed]
12. Brent, L.J.; Lehmann, J.; Ramos-Fernández, G. Social network analysis in the study of nonhuman primates: A historical perspective. *A. J. Primatol.* **2011**, *73*, 720–730. [CrossRef] [PubMed]
13. Wey, T.; Blumstein, D.T.; Shen, W.; Jordán, F. Social network analysis of animal behaviour: A promising tool for the study of sociality. *Anim. Behav.* **2008**, *75*, 333–344. [CrossRef]
14. Sporns, O. The human connectome: A complex network. *Ann. N. Y. Acad. Sci.* **2011**, *1224*, 109–125. [CrossRef] [PubMed]
15. Newton-Fisher, N.E. *Animal Behaviour Pro: V1*; Apple: Canterbury, UK, 2012.
16. Martin, P.; Bateson, P.P.G.; Bateson, P. *Measuring Behaviour: An Introductory Guide*; Cambridge University Press: Cambridge, UK, 1993.
17. Altmann, J. Observational study of behavior: Sampling methods. *Behaviour* **1974**, *49*, 227–266. [CrossRef] [PubMed]
18. Shannon, P.; Markiel, A.; Ozier, O.; Baliga, N.S.; Wang, J.T.; Ramage, D.; Amin, N.; Schwikowski, B.; Ideker, T. Cytoscape: A software environment for integrated models of biomolecular interaction networks. *Genome Res.* **2003**, *13*, 2498–2504. [CrossRef] [PubMed]
19. Kamada, T.; Kawai, S. An algorithm for drawing general undirected graphs. *Inf. Process. Lett.* **1989**, *31*, 7–15. [CrossRef]
20. Kerth, G.; Perony, N.; Schweitzer, F. Bats are able to maintain long-term social relationships despite the high fission–fusion dynamics of their groups. *Proc. R. Soc. Lond. B. Biol. Sci.* **2011**, *278*, 2761–2767. [CrossRef] [PubMed]
21. Voelkl, B.; Kasper, C.; Schwab, C. Network measures for dyadic interactions: Stability and reliability. *Am. J. Primatol.* **2011**, *73*, 731–740. [CrossRef] [PubMed]
22. Wrangham, R.; Crofoot, M.; Lundy, R.; Gilby, I. Use of overlap zones among group-living primates: A test of the risk hypothesis. *Behaviour* **2007**, *144*, 1599–1619. [CrossRef]
23. Hill, R.; Lee, P. Predation risk as an influence on group size in cercopithecoid primates: Implications for social structure. *J. Zool.* **1998**, *245*, 447–456. [CrossRef]
24. Jernvall, J.; Wright, P.C. Diversity components of impending primate extinctions. *Proc. Natl. Acad. Sci. USA* **1998**, *95*, 11279–11283. [CrossRef] [PubMed]

25. Rebecchini, L.; Schaffner, C.M.; Aureli, F. Risk is a component of social relationships in spider monkeys. *Ethology* **2011**, *117*, 691–699. [CrossRef]

26. Fedigan, L.M.; Baxter, M.J. Sex differences and social organization in free-ranging spider monkeys (*Ateles geoffroyi*). *Primates* **1984**, *25*, 279–294. [CrossRef]

27. Link, A.; Milich, K.; Di Fiore, A. Demography and life history of a group of white-bellied spider monkeys (*Ateles belzebuth*) in western amazonia. *Am. J. Primatol.* **2018**, e22899. [CrossRef] [PubMed]

28. Campbell, J.C.; Baty, M.; Ghandour, R.M.; Stockman, J.K.; Francisco, L.; Wagman, J. The intersection of intimate partner violence against women and hiv/aids: A review. *Int. J. Inj. Control Saf. Promot.* **2008**, *15*, 221–231. [CrossRef] [PubMed]

29. Jewkes, R.K.; Levin, J.B.; Penn-Kekana, L.A. Gender inequalities, intimate partner violence and hiv preventive practices: Findings of a south african cross-sectional study. *Soc. Sci. Med.* **2003**, *56*, 125–134. [CrossRef]

30. Güntürkün, O. Human behaviour: Adult persistence of head-turning asymmetry. *Nature* **2003**, *421*, 711. [CrossRef] [PubMed]

31. Chapelain, A.; Pimbert, P.; Aube, L.; Perrocheau, O.; Debunne, G.; Bellido, A.; Blois-Heulin, C. Can population-level laterality stem from social pressures? Evidence from cheek kissing in humans. *PLoS ONE* **2015**, *10*, e0124477. [CrossRef] [PubMed]

32. Ocklenburg, S.; Güntürkün, O. Head-turning asymmetries during kissing and their association with lateral preference. *Laterality* **2009**, *14*, 79–85. [CrossRef] [PubMed]

33. Sedgewick, J.R.; Elias, L.J. Family matters: Directionality of turning bias while kissing is modulated by context. *Laterality Asymmetries Body Brain Cogn.* **2016**, *21*, 662–671. [CrossRef] [PubMed]

34. Van der Kamp, J.; Canal-Bruland, R. Kissing right? On the consistency of the head-turning bias in kissing. *Laterality* **2011**, *16*, 257–267. [CrossRef] [PubMed]

35. Packheiser, J.; Rook, N.; Dursun, Z.; Mesenhöller, J.; Wenglorz, A.; Güntürkün, O.; Ocklenburg, S. Embracing your emotions: Affective state impacts lateralisation of human embraces. *Psychol. Res.* **2018**, 1–11. [CrossRef] [PubMed]

36. Turnbull, O.; Stein, L.; Lucas, M. Lateral preferences in adult embracing: A test of the "hemispheric asymmetry" theory of infant cradling. *J. Genet. Psychol.* **1995**, *156*, 17–21. [CrossRef]

symmetry

MDPI

Article

Meta-Control in Pigeons (*Columba livia*) and the Role of the Commissura Anterior

Emre Ünver [1,*], Qian Xiao [1,2] and Onur Güntürkün [1]

[1]   Department of Biopsychology, Institute of Cognitive Neuroscience, Faculty of Psychology,
     Ruhr University Bochum, 44801 Bochum, Germany; qianxiao@moon.ibp.ac.cn (Q.X.);
     Onur.Guentuerkuen@ruhr-uni-bochum.de (O.G.)
[2]   Key Laboratory of Interdisciplinary Science, Institute of Biophysics, Chinese Academy of Sciences,
     Beijing 100101, China
*   Correspondence: emre.uenver@rub.de; Tel.: +49-234-32-26213

Received: 30 November 2018; Accepted: 16 January 2019; Published: 22 January 2019

Abstract: Meta-control describes an interhemispheric response conflict that results from the perception of stimuli that elicit a different reaction in each hemisphere. The dominant hemisphere for the perceived stimulus class often wins this competition. There is evidence from pigeons that meta-control results from interhemispheric response conflicts that prolong reaction time when the animal is confronted with conflicting information. However, recent evidence in pigeons also makes it likely that the dominant hemisphere can slow down the subdominant hemisphere, such that meta-control could instead result from the interhemispheric speed differences. Since both explanations make different predictions for the effect of commissurotomy, we tested pigeons in a meta-control task both before and after transection of the commissura anterior. This fiber pathway is the largest pallial commissura of the avian brain. The results revealed a transient phase in which meta-control possibly resulted from interhemispheric response conflicts. In subsequent sessions and after commissurotomy, however, the results suggest interhemispheric speed differences as a basis for meta-control. Furthermore, they reveal that meta-control is modified by interhemispheric transmission via the commissura anterior, although it does not seem to depend on it.

Keywords: birds; hemispheric interactions; brain asymmetry; reaction time; color discrimination

---

1. Introduction

Meta-control refers to the one hemisphere taking charge of response selection when the two hemispheres are brought into conflict [1–3]. This phenomenon was first demonstrated in split-brain patients and healthy people [1,4], but was also later revealed in monkeys [5], chicken [6], and pigeons [2,3,7]. It is often assumed that meta-control results from one hemisphere inhibiting the other via the various commissures that connect the two halves of the brain at the midbrain and telencephalic level [8,9].

Meta-control becomes especially visible in species with pronounced brain asymmetries. Depending on the type of stimulus, one or the other hemisphere regularly gains control. Birds are ideal subjects for these studies [10]. Their left hemisphere is superior in discrimination, categorization, and memorization of visual patterns (chicks: [11]; quail: [12]; pigeons: [13,14]) and visuomagnetic cues (pigeons: [15]; chicks: [16]), while their right hemisphere is superior in visually guided interactions with emotionally charged stimuli (chicks: [17]), attentional shifts (chicks and pigeons: [18]), social interactions (chicks: [19]), as well as in relational and spatial analyses of visual information (chicks: [20]; pigeons: [14,21]).

Meta-control could result from either inter-hemispheric response conflict or differences in hemisphere-specific speed. If inter-hemispheric response conflict was the cause, situations in which each half-brain competes to present a different response should produce longer reaction times than non-conflicting situations [2,8]. This is because decision making with two incompatible options usually requires a longer processing time [10]. If, however, meta-control simply results from hemisphere-specific processing speed, the outcome would be different. The decision time would be determined solely by the faster hemisphere, which would always win. Two competing hemispheres would then be as fast as the faster hemisphere.

A recent study conducted by Ünver & Güntürkün [2] in pigeons collected evidence for the inter-hemispheric response conflict model. In their study, pigeons were trained by a forced-choice color discrimination task monocularly, and each hemisphere learned to discriminate between its own stimulus pair. Then, under binocular conditions, the birds were exposed to two types of test stimuli. These test stimuli were created by combining positive and negative patterns learned by each hemisphere. If the animal had to discriminate between a stimulus pair that consisted of two positive (left- and right-hemispheric) patterns on one pecking key and two negative patterns on the other, the choice was easy. Both hemispheres agreed to peck the pattern combination that was positive for both half-brains. Consequently, the animals responded quickly to this "super stimulus". The situation was different when each stimulus was composed of the positive pattern of one hemisphere and the negative pattern of the other hemisphere. In the case of such an "ambiguous stimulus", the overall pattern signaled an interhemispheric reward history conflict. As it turned out, the ambiguous stimulus caused a significant response delay. This makes it likely that meta-control rests mainly on an inter-hemispheric response conflict and not on hemisphere-specific speed.

A recent study, however, proposed a different mechanism. Qian Xiao & Güntürkün [22] recorded signals from the sensorimotor arcopallium of pigeons while the birds were conducting a color discrimination task under monocular conditions. All birds in their study learned faster and responded more quickly with their right eye/left hemisphere. The arcopallium not only harbors descending premotor neurons but also commissural neurons that constitute the commissura anterior—the largest avian interhemispheric connection at the pallial level. As shown by Letzner et al. [23], the commissura anterior originates from the telencephalic arcopallium/amygdala-complex and contains a small cluster of non-GABAergic sensorimotor and amygdaloid fibers that project onto a wide range of contralateral structures such as the posterior amygdala, the sensorimotor arcopallium, as well as further sensory and motor components of the nidopallium. We chose this commissure for our study due to these widespread projections onto the contralateral hemisphere. Xiao & Güntürkün [22] transiently blocked the arcopallial activity of one hemisphere and recorded from the contralateral arcopallium during color discrimination to determine the effect of left-to-right and right-to-left information transfer. They discovered that the left hemisphere was able to modify the timing of individual activity patterns of the neurons in the right hemisphere via asymmetrical commissural interactions. In contrast to that, right arcopallial neurons were hardly able to alter the activity pattern of left arcopallial cells. Thus, under conditions of interhemispheric competition, left arcopallial neurons could delay the contralateral spike time of those in the right hemisphere. As a result, the neurons of the right hemisphere would come too late to control a response and the left hemisphere would govern decisions. This finding could imply that hemispheric dominance in birds is realized at least in part by time shifts of the neural activity of one or the other hemisphere.

The studies by Ünver & Güntürkün [2] and Xiao & Güntürkün [22] make contradictory predictions of the mechanisms of meta-control. Both would assume that the commissura anterior plays a decisive role in inter-hemispheric response conflicts but would predict different choice patterns from birds in a meta-control task after commissurotomy. Ünver & Güntürkün [2] would infer that the loss of the commissura anterior should reduce reaction times when presented with an ambiguous stimulus because an inter-hemispheric response conflict could no longer result in an inter-hemispheric delay in processing time. In contrast, Xiao & Güntürkün [22] would not expect a change in reaction times under the ambiguous stimulus because the dominant hemisphere already determines the response. They would, however, expect that the dominance of the left hemisphere would weaken after commissurotomy because the left-to-right control of the neuronal spike times could no longer be executed. To test these predictions, we conducted a meta-control study as published by Ünver & Güntürkün [2], and subsequently transected the commissura anterior to re-test the animals with the same task.

2. Materials and Method

2.1. Subjects

Nine naïve pigeons of unknown sex were used in the study. All pigeons were housed in single cages with other conspecifics and maintained on a 12:12 h light–dark cycle. Their body weight was maintained at 80–90% of their free-feeding weight by feeding diet food on weekdays and a mixture of peas, corn, and sunflower seeds on the weekends. Water was provided ad libitum. For the monocular sessions, velcro rings were fixed around the eyes of the pigeons using glue that was non-irritating to the skin. Cone-shaped eye caps that were attached to the other sides of the velcro rings at their bases and were created using cardboard. These eye caps could be easily attached and removed from the rings surrounding the eyes for monocular testing (Figure 1). All procedures were conducted in compliance with the guidelines for the care and use of laboratory animals and approved by the local committee (LANUV).

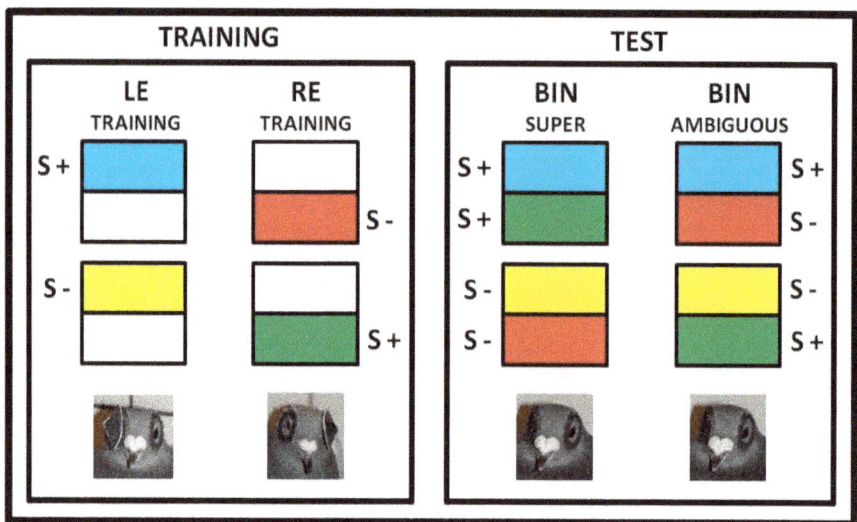

Figure 1. The stimuli used in the experiment. Super stimuli consisted of a combination of two positive and two negative stimuli presented to the left eye (LE) and right eye (RE) during the training phase. Ambiguous stimuli were created by combining a negative stimulus for one hemisphere and a positive stimulus for the other. Both eyes (BIN = binocular) were open during the test phase. The color combinations shown in the figure are merely examples of the various combinations used. Below are photographs showing the animals with a cap on one eye (**left**) or both eyes uncovered (**right**).

2.2. Apparatus

A custom-made operant chamber measuring 40 × 35 × 35 cm (W × D × H) in size was used for the experiment. The chamber was equipped with a feeder and illuminated using a house light. The feeder was immediately illuminated when food was presented. The stimuli (5 × 5 cm in size) were introduced on a TFT LCD touchscreen monitor with 1024 × 768 resolution. The monitor was placed on the same side of the chamber as the feeder to ensure that the pigeons could easily reach the feeder immediately after pecking at the stimuli on the screen. The experimental sessions were controlled by a custom-written MATLAB program (MathWorks, Natick, MA, USA) using the Biopsy Toolbox [24].

2.3. Procedure

Before learning the color discrimination task, all pigeons were trained in autoshaping sessions consisting of 40 trials. In these sessions, the pigeons were made to peck on a white square presented on the screen under monocular conditions. The white square was presented for 4 s, and food was delivered immediately following a single peck on the white square. These sessions were conducted according to a fixed ratio (FR1) schedule. The birds were trained in a counterbalanced manner—on one day, only the left eye (LE) was blocked, whereas on the next day, only the right eye (RE) was blocked. Response to the white square in >85% of the trials in two consecutive sessions per eye condition was set as the criterion for progress to the subsequent schedules. Once the birds met this criterion, their training progressed to a variable ratio (VR) schedule wherein they were progressively trained with variable ratios VR2, VR4, and VR8 under monocular conditions again, with the same criterion. All the sessions in the VR schedule consisted of 40 trials.

Once the birds met the response criterion for the VR, we commenced the color discrimination training. Rectangles of four different colors (red, yellow, green, or blue) were used as stimuli. The color discrimination sessions were conducted under monocular conditions, and the color combinations were balanced among pigeons to prevent color preferences. As shown in Figure 1, they were always placed in a compound at the upper or lower position of a larger white rectangle. Each eye of the pigeons was exposed to a different pair of stimuli (e.g., red and yellow for the LE; blue and green for the RE). One of these colors served as S+ and the other as S− for each eye. The pigeons had to choose between an upper and a lower compound stimulus that each consisted of a colored and a white rectangle. Pecks on the S+ compound were rewarded regardless of whether the peck location was on the colored or on the white part of the compound. The same rule was applied for the S− compound. The monocular sessions were conducted in a counterbalanced manner, similar to the autoshaping sessions.

The stimuli were presented for 4 s. A single peck on the S+ compound immediately activated the feeder for 2 s, whereas a peck on the S− compound resulted in switching off the house lights for 5 s and playing a loud noise for 1 s. Once the birds responded to the S+ compound in >85% of the trials in two consecutive sessions for each eye condition, the number of trials per session was increased to 200 in steps of 20. The criteria that was applied in each step was that the pigeons had to make at least 85% correct choices (responses to the S+ compound) for each eye condition in a single session. As the number of trials in each session was increased, the reward ratio (responses to S+) was decreased in steps of 10% until reaching 40%. This procedure was employed to prevent extinction learning in subsequent catch trials. As a final step, a new stimulus pair, a white (S+) square and a gray (S−) square, were introduced. Because the birds had already been trained to respond to the white square during the autoshaping sessions, we expected them to be able to rapidly discriminate between this new stimulus pair. This white/gray "dummy" discrimination procedure was necessary to maintain the birds' responses during the critical test sessions that included catch trials. In the catch trials, the colored stimuli were re-arranged to create "super" and "ambiguous" stimuli that were not rewarded. Each of the final sessions consisted of 200 trials, with 80% of the stimuli being presented as white (S+) and gray (S−) dummy stimuli. As outlined above, both S+ (the S+ of the LE and the S+ of the RE) on one pecking key and both S− on the other key were termed super stimuli. Unlike the other sessions, the critical test sessions were performed under binocular conditions. The gray/white stimuli represented a common associative background for both stimuli. This was not applied to the ambiguous stimuli. On each key, the S+ of one hemisphere was always combined with the S− of the other hemisphere. The proportion of catch trials in the final session was 20% (i.e., the number of catch trials was 40, with 20 being ambiguous and 20 being super stimuli). The remaining trials consisted of the white/gray stimuli pair (the number of white/gray stimuli was 160). No feedback for the catch trials was available, whereas the white/gray stimuli discrimination had a 40% reward probability. Following the first critical test session that included catch trials, the pigeons were further trained using the well-known training stimuli under monocular conditions. These sessions using the well-known training stimuli between each critical test session were conducted because it was necessary to maintain the pigeons' response at a stable level during the subsequent critical test sessions. Therefore, this sequence was repeated until enough catch trial responses were collected.

After six sessions at most of testing for meta-control, pigeons underwent a commissurotomy operation. After a two-week recovery period, the same task and procedure were applied, and data were collected.

2.4. Surgery

Before surgery, nine birds participating in the experiment were given a mixture of ketamine (ketamine hydrochloride, 100 mg/mL; Zoetis, Berlin, Germany) and xylazine (xylazine hydrochloride, 23.32 mg/mL, methyl-4-hydroxybenzoate, 1.5 mg/mL; Bayer Vital, Leverkusen, Germany) by intramuscular injection (7:3 ratios, 0.12 mL/100 g body weight). The anesthetized birds were placed on a warming pad in a stereotaxic device. Their heads were fixed at a 45° angle in the head holder according to the coordinates of the pigeon brain atlas [25]. Prior to the commissurotomy, the scalp was opened and a window was opened in the skull with a drill, centered at the anterior 7.75 and lateral 0.0 coordinates. Then, the dura mater was removed. The main vessel in the gap between the two hemispheres was delicately pulled aside with a hand-made hook. Finally, a 2-mm-wide, 0.3 mm thick blade was slowly lowered into the region with the following coordinates: Anterior 7.75, lateral 0.0 at a depth of 9.0 mm from the surface of the brain [25]. The blade was lowered in increments of 1 mm, with a 2 min pause between each increment. Thus, the risk of damage to the brain due to the pressure caused by the blade was minimized. At the end of the operation, the knife was removed in the same manner, i.e., by lifting 1 mm every 2 min. The skin was stitched after a medical sponge was placed on the operation area. Finally, a painkiller was sprayed over the operation area and an antibacterial powder (Tyrasor; Engelhard Arzneimittel, Niederdorfleben, Germany) was applied. In addition, an intramuscular painkiller (Rimadyl, 0.04 mL/100 g body weight; Pfizer, GmbH, Münster, Germany) was administered. The pigeons were kept in their individual cages for one week to allow them to overcome the effects of the operation. Then, the tests were conducted.

2.5. Histology

The pigeons were deeply anesthetized with equithesin (0.55 mL/100 g body weight) and perfused with 4% paraformaldehyde (VWR Prolabo Chemicals, Leuven, Belgium) after the last post-operation tests. The brain was removed, immersed in gelatin (Merck, Darmstadt, Germany) and sectioned into 40-μm frontal slices using a freezing microtome (Leica Microsystems Nussloch GmbH, Nussloch, Germany). Sections were mounted, nissl and klüver-barrera stained, and the success of the commissurotomy was verified microscopically. In all nine birds, the commissura anterior was verified to be completely sectioned (Figure 2). In some animals the blade had been successfully lowered along the midline (Figure 2b), in others it was slightly off the midline and had damaged the medial most parts of the hemispheres in the medial meso- and nidopallium, as well the area above the commissura anterior (Figure 2a). These are not areas associated with the visual system and we could not see any correlation between our histological verifications and our behavioral results.

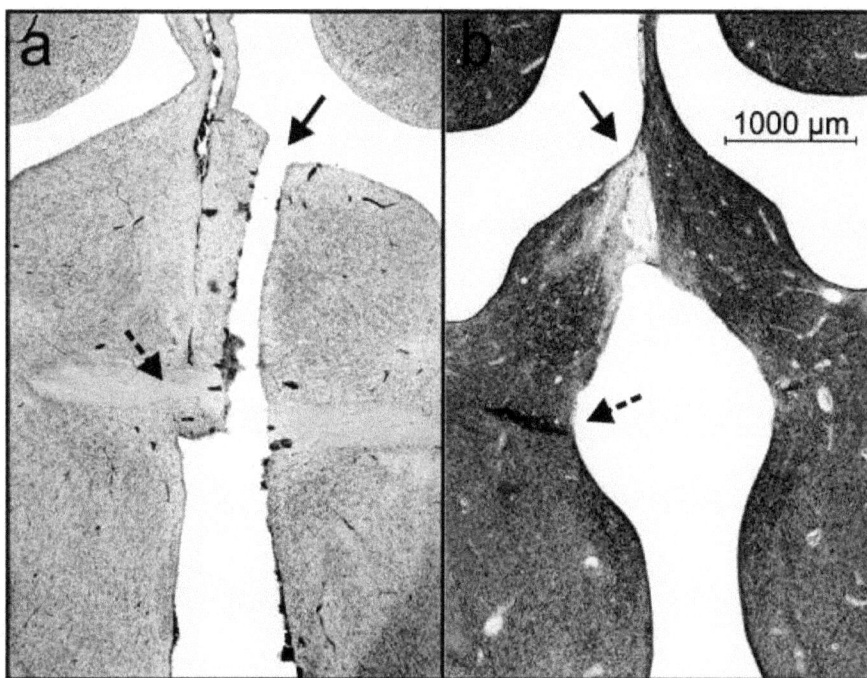

Figure 2. A nissl (**a**) and a nissl/klüver-barrera (**b**) stained frontal section of two pigeons with transections of the commissura anterior. The straight arrows point to the tissue rupture resulting from the passing of the blade, while the broken arrows indicate remaining fibers of the commissura. Note that in (**a**) the blade has damaged the area above the commissure since it was slightly off the midline. This is not the case in (**b**). Scale bar in (**b**) also applies to (**a**).

3. Results

Two variables were important in studying the effect of the commissurotomy on meta-control. First, how many individuals display significant meta-control before vs. after commissurotomy? Meta-control in our task is defined as a significantly higher number of choices that are dominated by one hemisphere being faced with an ambiguous pattern. Second, how did the reaction times to ambiguous- and super-stimuli change after the commissurotomy?

Meta-control: A meta-control effect was observed in three out of nine birds before commissurotomy (for each individual: chi square test, $p < 0.05$). In two birds the right eye dominated the decisions of the animal, and in one bird the left eye was dominant. Overall, this number was not sufficient to produce a significant meta-control effect at the population level (paired-sample t-test, $t = 0.246$, $p = 0.812$, $n = 9$). These three birds all ceased to demonstrate meta-control after commissurotomy. On the other hand, post-commissurotomy meta-control was observed in two different animals (one left, one right eye) that had not exhibited meta-control before the operation (chi square test, each $p < 0.05$). During the post-commissurotomy period, no significant meta-control at the population level was observed (paired-sample t-test, $t = 0.939$, $p = 0.375$, $n = 9$).

Reaction times: When first confronted with the ambiguous stimulus, the birds showed significantly higher reaction times to the ambiguous (1.14 s) than to the super stimulus (1.03 s) (paired-sample t-test, $t = 2.540$, $p = 0.035$, $n = 9$). In the second and subsequent sessions, however, this effect disappeared, such that the reaction time responses to super and ambiguous stimuli were no longer significantly different from each other (super stimulus: 1.07 s; ambiguous stimulus: 1.1 s; (paired-sample t-test, $t = 0.479$, $p = 0.646$, $n = 8$)). There were no significant reaction time differences to the super stimulus between session 1 and sessions 2–6 (paired sample t-test; $t = 0.755$, $p = 0.475$, $n = 8$). The same applied to the ambiguous stimulus (paired-sample t-test; $t = 0.033$, $p = 0.975$, $n = 8$). Note that the average values of sessions 2–6 were derived from 8 birds, since one pigeon stopped working on the task after session 1 (and then restarted after surgery). Similarly, in the post-surgery tests, no significant differences in the reaction times between super and ambiguous signals were observed (super stimulus: 1.24 s; ambiguous stimulus: 1.29 s; (paired-sample t-test, $t = 0.614$, $p = 0.556$, $n = 9$)). Moreover, there was no significant difference between the response times to the two stimulus types in the pre-surgery sessions (excluding session 1) and post-surgery sessions (mean of super stimulus sessions 2–6: 1.07 s; post-surgery session: 1.15 s; (paired-sample t-test, $t = 0.680$, $p = 0.518$, $n = 8$); mean of ambiguous stimulus sessions 2–6: 1.1 s; post-surgery session: 1.22 s; (paired-sample t-test, $t = 1.097$, $p = 0.309$, $n = 8$)) (Figure 3).

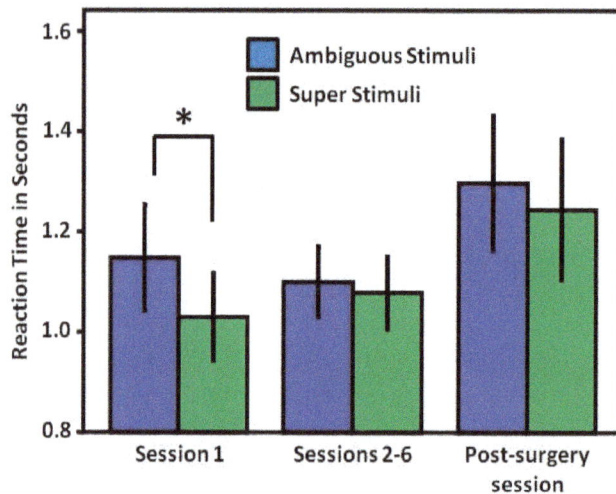

Figure 3. Average reaction times of subjects to ambiguous and super stimuli during sessions prior to the commissurotomy and in the first session after the commissurotomy. Significant differences are indicated by an asterisk (p <0.05). Error bars are ±1 SEM. Note that the averages of sessions 2–6 were derived from 8 birds, because one pigeon stopped working on the task after session 1, but restarted after surgery.

4. Discussion

Meta-control can occur when the two hemispheres compete with each other to produce a hemisphere-specific response [1,2,4,5,7]. In studies with birds working on color discrimination tasks, the dominant hemisphere is usually the left [10,11,13]. Concomitantly, there is some evidence for a higher incidence of left-hemispheric meta-control in such tasks with pigeons [7]. The present study tested two different possible mechanisms of meta-control. One of these assumes that meta-control results from each hemisphere inhibiting the other [8]. Such a mechanism should cause conflicting (in our case ambiguous) stimuli to produce longer processing times, resulting in longer reaction times. A recent study found evidence supporting this prediction, and therefore suggested that meta-control results from the interhemispheric conflict [2]. An electrophysiological study, however, found evidence for a different mechanism: Xiao & Güntürkün [22] discovered that arcopallial neurons of the left hemisphere dominate the response of the animal during color discrimination through a faster activation of motor responses. Furthermore, the left hemisphere controls the right hemispheric spike times, and is thus able to delay reaction times of the other hemisphere. This effect would increase the advantage of the left hemisphere. These findings make different predictions for the effect of the commissurotomy on meta-control. The mechanism based on the interhemispheric conflict would imply that a section of the commissura anterior should reduce reaction times to ambiguous stimuli (no commissural exchange → no interhemispheric conflict), whereas the model based on hemisphere specific speed would not predict post-surgery changes in reaction time to ambiguous stimuli (no commissural exchange → no change in hemisphere-specific speed). At the same time, the results of Xiao & Güntürkün [22] suggest that the advantage of the left hemisphere would be smaller after commissurotomy (no commissural exchange → no possibility to further delay response execution of the right hemisphere). Our findings suggest that the birds only experience interhemispheric conflict on the first session with ambiguous stimuli, and the effect disappears in the following sessions. A subsequent commissurotomy does not alter reaction times to ambiguous stimuli but does modify meta-control. Overall, our data would be compatible with a model according to which interhemispheric conflict occurs in a short, initial period, but then gives way to lateralized reaction patterns determined by hemisphere-specific speed.

As visible in Figure 3, reaction times to super and ambiguous stimuli were the most different in the first session in which the animals were first presented these two stimulus types under binocular conditions. However, in subsequent sessions reaction times became increasingly similar. Ünver & Güntürkün [2] had based their conclusion of interhemispheric conflict on the first session after introducing ambiguous stimuli. This conclusion may remain valid but is obviously restricted to this initial session. In subsequent sessions, a different mechanism seems to prevail. It is indeed conceivable that the animals quickly learned about the absence of negative or positive feedback when responding to the ambiguous stimuli. It is known that pigeons are extremely sensitive to reward alterations in operant categorization tasks, and subsequently tend to bias their choices towards initially favored alternatives [26]. Similar findings were also observed in studies with monkeys [27,28]. This makes it likely that our commissurotomy was performed at a point in time in which the pigeons were no longer pondering response conflicts but instead biased their choices according to mechanisms based on hemisphere-specific speed. Consequently, response times to ambiguous stimuli were not altered by commissurotomy.

This scenario is compatible with the explanation that each hemisphere rushes with its own hemisphere-specific speed to motor areas. During color discrimination, the left hemisphere usually produces faster reaction times. This has been observed in various studies with pigeons [29] and other birds [17,30]. This was also observed by Xiao & Güntürkün [22] when recording from the pigeon arcopallium during color discrimination. This study also offers a mechanistic explanation of this observation by revealing that the left hemisphere can modify the spike time of the right hemisphere. Thus, under conditions of conflict, the left hemisphere could delay the right hemispheric response speed, thereby accelerating its own advantage. From this point of view, a transection of the commissura anterior should reduce, but not completely terminate the left hemispheric superiority. Indeed, we observed major alterations of meta-control after surgery. Usually, an individually significant extent of meta-control is observed in only a fraction of pigeons [2,3,7]. With the procedure used in this study, it was mostly the left hemisphere that evinced meta-control [2,7]. In the current experiment, three out of nine birds demonstrated meta-control before commissurotomy (two left hemispheric, one right hemispheric). This is a typical result pattern [2,7]. After transecting the commissura anterior, however, all three birds lost their hemisphere-specific advantage. Instead, two other birds displayed significant meta-control (one left, one right). Although this is certainly not a strong proof of the conclusion of Xiao & Güntürkün [22], it is conceivable that the changes observed in meta-control in our nine pigeons resulted from the loss of a left hemispheric advantage that resulted in biased interhemispheric interactions. If indeed neuronal speed differences cause the bias towards the right eye in metacontrol studies, the large individual differences may result from the fact that neurons show within the pigeon's visual system substantial latency differences between individual birds [22,31–33].

It is known that the commissura anterior connects with the anterior and intermediate arcopallium. These structures project onto a wide cluster of visual and sensorimotor areas. Our study focused on the contribution of the commissura anterior to visual asymmetries. However, further commissural systems may also play a role in metacontrol since studies of both chicks [34] and pigeons [35–37] suggested that subpallial commissures also play key roles in visually-guided lateralized behavior. The supraoptic decussation (DSO) is one such subpallial connection, and is known to be responsible for interocular transfer during visual discrimination [38]. This may be due to the indirect connection of the DSO to telencephalic visual structures such as Wulst. More recently, it has been shown that the nucleus of the lateral ponto-mesencephalic tectum (nLPT), a midbrain structure, contains GABAergic neurons and its projections terminate in the contralateral optic tectum (TeO) via the commissura tectalis [39]. Therefore, this midbrain commissure may also play a crucial role during meta-control. Thus, the present study must be complemented by further experiments to reveal the full scenario of interhemispheric interactions of lop-sided bird brains.

Although our study was centered on the mechanisms of meta-control, it might also offer some more general insights on the behavior of organisms with lateralized brains. A key problem of these species is the production of a single response from two asymmetrically specialized hemispheres. Our results suggest that the default option in such situations could be to let both hemispheres compete based on hemisphere-specific processing speed. Because the dominant hemisphere for a certain stimulus class usually produces faster responses [22], the most competent half-brain would primarily determine the response. The commissural slowing mechanism discovered by Xiao & Güntürkün [22] would amplify this interhemispheric speed difference to ensure that the dominant hemisphere controls the overall response.

Author Contributions: Conceptualization: Q.X. and O.G.; designed experiment: Q.X. and O.G.; performed experiment: E.Ü.; statistical analysis: E.Ü. and O.G.; manuscript preparation: E.Ü. and O.G.; funding acquisition and project supervision: O.G. All authors revised and approved the paper.

Acknowledgments: We are grateful for the support of Annika Simon during surgery and the conduct of the histological procedure. We also thank Felix Ströckens and Sarah von Eugen for help during documentation of histological results. Supported by the Deutsche Forschungsgemeinschaft through SFB 874.

Conflicts of Interest: The authors declare no conflict of interest.

References

1. Levy, J.; Trevarthen, C. Metacontrol of hemispheric function in human split-brain patients. *J. Exp. Psychol. Hum.* **1976**, *2*, 299–312. [CrossRef]
2. Ünver, E.; Güntürkün, O. Evidence for interhemispheric conflict during meta-control in pigeons. *Behav. Brain Res.* **2014**, *270*, 146–150. [CrossRef] [PubMed]
3. Adam, R.; Güntürkün, O. When one hemisphere takes control: Metacontrol in pigeons (*Columba livia*). *PLoS ONE* **2009**, *4*, e5307. [CrossRef] [PubMed]
4. Urgesi, C.; Bricolo, E.; Aglioti, S.M. Hemispheric metacontrol and cerebral dominance in healthy individuals investigated by means of chimeric faces. *Cogn. Brain Res.* **2005**, *24*, 513–525. [CrossRef] [PubMed]
5. Kavcic, V.; Fei, R.; Hu, S.; Doty, R.W. Hemispheric interaction, meta-control, and mnemonic processing in split-brain macaques. *Behav. Brain Res.* **2000**, *111*, 71–82. [CrossRef]
6. Vallortigara, G. Comparative neuropsychology of the dual brain: A stroll through animals' left and right perceptual worlds. *Brain Lang.* **2000**, *73*, 189–219. [CrossRef]
7. Freund, N.; Valencia-Alfonso, C.E.; Kirsch, J.; Brodmann, K.; Manns, M.; Güntürkün, O. Asymmetric top-down modulation of ascending visual pathways in pigeons. *Neuropsychologia* **2016**, *83*, 37–47. [CrossRef] [PubMed]
8. Chiarello, C.; Maxfield, L. Varieties of interhemispheric inhibition, or how to keep a good hemisphere down. *Brain Cogn.* **1996**, *30*, 81–108. [CrossRef]
9. Zeier, H.J.; Karten, H.J. Connections of the anterior commissure in the pigeon (*Columba livia*). *J. Comp. Neurol.* **1973**, *150*, 201–216. [CrossRef]
10. Rogers, L.J.; Vallortigara, G.; Andrew, R.J. *Divided Brains: The Biology and Behaviour of Brain Asymmetries*; Cambridge University Press: Cambridge, UK, 2013.
11. Rogers, L.J. Asymmetry of brain and behavior in animals: Its development, function, and human relevance. *Genesis* **2014**, *52*, 555–571. [CrossRef]

12. Valenti, A.; Sovrano, V.A.; Zucca, P.; Vallortigara, G. Visual lateralisation in quails (*Coturnix coturnix japonica*). *Laterality* **2003**, *8*, 67–78. [CrossRef] [PubMed]

13. Güntürkün, O.; Kesch, S. Visual lateralization during feeding in pigeons. *Behav. Neurosci.* **1987**, *101*, 433–435. [CrossRef] [PubMed]

14. Yamazaki, Y.; Aust, U.; Huber, L.; Hausmann, M.; Güntürkün, O. Lateralized cognition: Asymmetrical and complementary strategies of pigeons during discrimination of "human concept". *Cognition* **2007**, *104*, 315–344. [CrossRef] [PubMed]

15. Prior, H.; Wiltschko, R.; Stapput, K.; Güntürkün, O.; Wiltschko, W. Visual lateralization and homing in pigeons. *Behav. Brain Res.* **2004**, *154*, 301–310. [CrossRef]

16. Rogers, L.J.; Munro, U.; Freire, R.; Wiltschko, R.; Wiltschko, W. Lateralized response of chicks to magnetic cues. *Behav. Brain Res.* **2008**, *186*, 66–71. [CrossRef]

17. Rogers, L.J. Development and function of lateralization in the avian brain. *Brain Res. Bull.* **2008**, *76*, 235–244. [CrossRef]

18. Diekamp, B.; Regolin, L.; Güntürkün, O.; Vallortigara, G. A left-sided visuospatial bias in birds. *Curr. Biol.* **2005**, *15*, R372–R373. [CrossRef]

19. Vallortigara, G.; Andrew, R.J. Differential involvement of right and left hemisphere in individual recognition in the domestic chick. *Behav. Proc.* **1994**, *33*, 41–58. [CrossRef]

20. Vallortigara, G.; Pagni, P.; Sovrano, V.A. Separate geometric and non-geometric modules for spatial reorientation: Evidence from a lopsided animal brain. *J. Cogn. Neurosci.* **2004**, *16*, 390–400. [CrossRef]

21. Pollonara, E.; Guilford, T.; Rossi, M.; Bingman, V.P.; Gagliardo, A. Right hemisphere advantage in the development of route fidelity in homing pigeons. *Anim. Behav.* **2017**, *123*, 395–409. [CrossRef]

22. Xiao, Q.; Güntürkün, O. Asymmetrical commissural control of the subdominant hemisphere in pigeons. *Cell Rep.* **2018**, *25*, 1171–1180. [CrossRef] [PubMed]

23. Letzner, S.; Simon, A.; Güntürkün, O. Connectivity and neurochemistry of the commissura anterior of the pigeon (*Columba livia*). *J. Comp. Neurol.* **2016**, *524*, 343–361. [CrossRef] [PubMed]

24. Rose, J.; Otto, T.; Dittrich, L. The Biopsychology-Toolbox: A free, open-source Matlab-toolbox for the control of behavioral experiments. *J. Neurosci. Meth.* **2008**, *175*, 104–107. [CrossRef] [PubMed]

25. Karten, H.J.; Hodos, W. *A Stereotaxic Atlas of the Brain of the Pigeon: Columba Livia*; Johns Hopkins Press: Baltimore, MD, USA, 1967.

26. Stüttgen, M.; Yildiz, A.; Güntürkün, O. Adaptive criterion setting in perceptual decision making. *J. Exp. Anal. Behav.* **2011**, *96*, 155–176. [CrossRef] [PubMed]

27. Feng, S.; Holmes, P.; Rorie, A.; Newsome, W.T. Can monkeys choose optimally when faced with noisy stimuli and unequal rewards? *PLoS Comput. Biol.* **2009**, *5*, e1000284. [CrossRef] [PubMed]

28. Teichert, T.; Ferrara, V.P. Suboptimal integration of reward magnitude and prior reward likelihood in categorical decisions by monkeys. *Front. Neurosci.* **2010**, *4*, 1–13. [CrossRef] [PubMed]

29. Güntürkün, O. Lateralization of visually controlled behavior in pigeons. *Physiol. Behav.* **1985**, *34*, 575–577. [CrossRef]

30. Güntürkün, O. Avian visual lateralization: A review. *Neuroreport* **1997**, *8*, 3–11.

31. Verhaal, J.; Kirsch, J.A.; Vlachos, I.; Manns, M.; Güntürkün, O. Lateralized reward-associated visual discrimination in the avian entopallium. *Eur. J. Neurosci.* **2012**, *35*, 1337–1343. [CrossRef]

32. Folta, K.; Troje, N.; Güntürkün, O. Timing of ascending and descending visual signals predicts the response mode of single cells in the thalamic nucleus rotundus of the pigeon (*Columba livia*). *Brain Res.* **2007**, *1132*, 100–109. [CrossRef]

33. Folta, K.; Diekamp, B.; Güntürkün, O. Asymmetrical modes of visual bottom-up and top-down integration in the thalamic nucleus rotundus of pigeons. *J. Neurosci.* **2004**, *24*, 9475–9485. [CrossRef] [PubMed]

34. Parsons, C.H.; Rogers, L.J. Role of the tectal and posterior commissures in lateralization of the avian brain. *Behav. Brain Res.* **1993**, *54*, 153–164. [CrossRef]

35. Güntürkün, O.; Böhringer, P.G. Lateralization reversal after intertectal commissurotomy in the pigeon. *Brain Res.* **1987**, *408*, 1–5. [CrossRef]

36. Skiba, M.; Diekamp, B.; Prior, H.; Güntürkün, O. Lateralized interhemispheric transfer of color cues: Evidence for dynamic coding principles of visual lateralization in pigeons. *Brain Lang.* **2000**, *73*, 254–273. [CrossRef] [PubMed]

37. Keysers, C.; Diekamp, B.; Güntürkün, O. Evidence for physiological asymmetres in the phasic intertectal interactions in the pigeon (*Columba livia*) and their potential role in brain lateralisation. *Brain. Res.* **2000**, *852*, 406–413. [CrossRef]
38. Watanabe, S. Interhemispheric transfer of visual discrimination in pigeons with supraoptic decussation (DSO) lesions before and after monocular learning. *Behav. Brain Res.* **1985**, *17*, 163–170. [CrossRef]
39. Stacho, M.; Letzner, S.; Theiss, C.; Manns, M.; Güntürkün, O. A GABAergic tecto-tegmento-tectal pathway in pigeons. *J. Comp. Neurol.* **2016**, *524*, 2886–2913. [CrossRef]

symmetry

MDPI

Article

Lateral Asymmetry of Brain and Behaviour in the Zebra Finch, *Taeniopygia guttata*

Lesley J. Rogers *, Adam Koboroff and Gisela Kaplan

School of Science and Technology, University of New England, Armidale, NSW 2351, Australia;
akoboroff@gmail.com (A.K.); gkaplan@une.edu.au (G.K.)
* Correspondence: lrogers@une.edu.au; Tel.: +61-266-515-006

Received: 1 November 2018; Accepted: 29 November 2018; Published: 1 December 2018

Abstract: Lateralisation of eye use indicates differential specialisation of the brain hemispheres. We tested eye use by zebra finches to view a model predator, a monitor lizard, and compared this to eye use to view a non-threatening visual stimulus, a jar. We used a modified method of scoring eye preference of zebra finches, since they often alternate fixation of a stimulus with the lateral, monocular visual field of one eye and then the other, known as biocular alternating fixation. We found a significant and consistent preference to view the lizard using the left lateral visual field, and no significant eye preference to view the jar. This finding is consistent with specialisation of the left eye system, and right hemisphere, to attend and respond to predators, as found in two other avian species and also in non-avian vertebrates. Our results were considered together with hemispheric differences in the zebra finch for processing, producing, and learning song, and with evidence of right-eye preference in visual searching and courtship behaviour. We conclude that the zebra finch brain has the same general pattern of asymmetry for visual processing as found in other vertebrates and suggest that, contrary to earlier indications from research on lateralisation of song, this may also be the case for auditory processing.

Keywords: asymmetry of brain function; lateralised behaviour; song; songbirds; zebra finch; predator inspection; eye preference; hemisphere differences; monocular viewing; general pattern of lateralisation

1. Introduction

It is timely to bring together and discuss the evidence for asymmetry of brain function in the zebra finch for two reasons. Firstly, the zebra finch is a model species used frequently to understand the links between neural structure and behaviour. Secondly, early research reporting lateral asymmetries in the species was equivocal, largely because it seemed to be at odds with lateralities reported in other avian species and because results of different studies were not always consistent. Therefore, we decided to summarise the available literature showing, or not showing, lateralisation in the zebra finch and to add some data on eye preference to view a predator.

The zebra finch has featured amongst those songbirds investigated for song learning, song production, and perception. Zebra finch song is stereotyped and has a rich spectro–temporal structure, which some researchers have compared to human speech sounds [1]. Furthermore, male zebra finches learn their song from other birds, by imitating the song of a tutor heard during a sensitive period of development [2,3]. These and other aspects of zebra finch song have been studied in considerable detail and compared to speech in humans [4–7].

Another feature of song is differential control of its production and processing by the left and right hemispheres. This has been studied in number of avian species, and studies of species other than the zebra finch have demonstrated a dominant role of song centres in the left hemisphere for controlling song production [8,9] and differential roles of the hemispheres in perception of song [10]. However, lateralisation of the song system in the zebra finch seemed not to fit this pattern.

Initially, Nottebohm et al. [10] cited unpublished observations that indicated little hemispheric asymmetry of song control in the zebra finch, and differing from the canary, zebra finches were reported to have no asymmetry in the size of the left and right hypoglossal nuclei, i.e., the collections of cell bodies with axons that form the hypoglossal nerves, a branch of which innervates the syringeal muscles used to produce song [11]. However, later research revealed the presence of asymmetry for song in the zebra finch, albeit not the same as that found in other passerine species.

Williams et al. [12] found right hemispheric control of song production in zebra finches; opposite to the direction of asymmetry reported for other songbirds. Lesioning the auditory areas of the right hemisphere of zebra finches was found to decrease the birds' ability to process harmonic structure in song [13]. Floody and Arnold [14] also reported evidence that the right song system is dominant in the zebra finch. Using functional magnetic resonance imaging (fMRI), Voss et al. [15] revealed hemispheric asymmetry in neural activity during stimulation by song: significant discrimination between songs was found only in the right hemisphere. Recognition of the zebra finch's own song versus the song of a conspecific was also found to be biased to the right hemisphere [16]. All of these studies indicated that perceptual production and processing was a function of the right hemisphere in zebra finches, and thus the asymmetry seemed to be reversed compared to other songbird species studied. However, measuring expression of the immediate early gene ZENK in zebra finches exposed to the auditory and/or visual aspects of courtship, Avery et al. [17] found left hemispheric dominance (i.e., hearing courtship song and seeing dancing by the courting male causes more neural activity in the left than the right hemisphere). Recent studies have demonstrated that both hemispheres attend to song but to different aspects of it [1], and that the direction of asymmetry depends on whether the memory of song is old or new [6].

It is possible that variation in the direction of asymmetry occurs depending on previous exposure to song and to what extent the birds recalled their previous exposure to song. Demonstrating that the direction of lateralisation depends on learning and memory, Moorman et al. [18] reported left-sided dominance of ZENK expression in the higher vocal centre of juvenile male zebra finches exposed to their tutor's song but not in those exposed to unfamiliar song. Olson et al. [6] found that the direction of laterality of song memory depends on strength of learning; the more the zebra finches learnt and remembered the song of their first tutor, the more right lateralised they were, as assessed by ZENK expression. By contrast, the more they learnt from a second tutor, the more left-lateralised they were. Hence, new and old memories of song appear to be located in opposite hemispheres; older memories in the right hemisphere and newer memories in the left hemisphere. In fact, Yang and Vicario [7] showed that exposure of adult zebra finches to novel hetero-specific sounds (vocalisations of canaries) can shift lateralisation for song processing from the right to the left hemisphere.

Using fMRI measurements of neural processing of song in zebra finches, Van Ruijevelt et al. [1] provided evidence that the spectral aspect of song is processed in the right hemisphere. By comparison, presentation of song with the spectral component filtered out, but with the temporal component remaining, led to greater neural activity in the left hemisphere [1]. Hence, the left hemisphere processes the temporal domain of song, whereas the right hemisphere processes the spectral component of song. This role of the left hemisphere in processing temporal aspects of song is supported by finding higher expression of ZENK in regions of the left hemisphere in males when they responded to arrhythmic, but not rhythmic, song [19].

A female zebra finch hearing a male's song directed towards her (as compared to the song produced by the male when he is alone) expresses a higher level of activity in the caudocentral–nidopallial region of the left hemisphere and the caudomedial–mesopallial region of the right hemisphere, as shown by functional magnetic imaging (fMRI) and early gene expression [20]. This result demonstrates that both hemispheres respond to hearing the song, but in each hemisphere the information is processed in different regions.

None of the above studies considered a possible role of visual asymmetry in association with lateralised auditory perception, learning, and production of song. Although it was known that fledgling zebra finches learn their tutor's song only when they can see the tutor [21], possible eye preference and lateralisation was not studied at this time. It was in the late 1970s that asymmetry of visual behaviour in an avian species was discovered (i.e., in the domestic chick [22]). Almost a decade later, the first asymmetry recorded for visual behaviour of the zebra finch concerned courtship. Workman and Andrew [23] reported evidence that during courtship males show a preference to use the right eye when viewing their female. This asymmetry was found by measuring the approach of the male to his female partner as he moved along a perch. Using a different method (viz., direction of movement in a circular corridor around cages containing females), Ten Cate et al. [24] failed to find any evidence for asymmetry, a result which Workman and Andrew [25] attributed to the males being tested with females that were not necessarily their own partners (see reply by Ten Cate [26]). A subsequent study by George et al. [27] resolved this debate by measuring brain activity during courtship singing by male zebra finches, and by measuring the amount of singing when only the left or right eye could be used. Males sang more song motifs when they could see a female with their right eye than they did when they could see her with their left eye. However, birds tested binocularly sang with more motifs than either of these groups.

Right eye preference during courtship approach to the female and in producing song with more motifs implies that this behaviour depends on the left hemisphere, since inputs from the eyes are largely processed by the contralateral hemisphere. In the study by George et al. [27], brain activity during courtship singing was assessed by assaying the expression of the immediate early genes, egr-1 and c-fos, in the optic tecta. The optic tectum on each side of the brain is the first relay station for visual inputs from the retina, and each eye sends its inputs to the contralateral optic tectum [28]. In males able to view the female with both eyes, neural activity was found to be higher in the left than the right optic tectum [27], and also in other regions of the left hemisphere [29]. This result is consistent with the preference to use the right eye during courtship singing, as found by Workman and Andrew [23]. It might be explained by the ability of the left hemisphere (and right eye) to sustain attention on a preferred and familiar stimulus [30–32] and to maintain attention on a stimulus towards which a motor response is planned [27].

A right eye preference during courtship by zebra finch males was confirmed by testing birds with monocular eye patches [33]. Birds that could see with their right eye only courted females more than those using their left eye only. The former also expressed preferences for orange-beaked (high quality) females over grey-beaked (low quality) females, whereas birds using their left eye only expressed no such preference [33].

Asymmetry in the motor behaviour of the zebra finch has also been reported. Alonso [34] measured side biases in allopreening and bill wiping and found significant preference to turn to the right side of the body for both of these responses. Since no left-right bias had been found for turning in a Y-maze, Alonso [34] interpreted her results as reflecting visual and not motor asymmetry, arguing that allopreening demands visual precision or that there is a need to keep the bird being preened in the right visual field. In the case of bill wiping, it was argued that the right-side preference reflects a right eye preference for pecking at food.

The zebra finch also uses the right eye and left hemisphere when it pecks at grain scattered amongst distracting pebbles. Alonso [35] found zebra finches tested with monocular eye patches could distinguish grain from pebbles when they were able to use their right eye only but not when they were able to use their left eye only [35]. This asymmetry reflected the role of the left hemisphere in performing this task. Indeed, this result replicated earlier research on domestic chicks, which had shown specialisation of the left hemisphere for learning a pebble-grain discrimination task [22,36], found also in pigeons [37,38].

Since attention and response to predators have been shown to be a specialisation of the right hemisphere, and left eye, in other avian species (domestic chicks [30], Australian magpies [39,40], as well as in amphibian [41] and mammalian [42] species), it seemed appropriate to test zebra finches for possible lateralisation of attention to a predator. Lombardi and Curio [43] have described the zebra finch's response to seeing a live owl as including side to side movements of the head which allow monocular fixation. Fixation refers to holding the head still after a rapid movement of the head.

We therefore needed to assess not only single monocular fixations of a stimulus, as in the species tested so far, but also biocular alternating fixations. Biocular alternating fixation refers to swapping from the lateral field of one eye to the lateral field of the other eye, as defined by Butler et al. [44], and it is distinguished from binocular fixation that uses the frontal visual field of both eyes.

Zebra finches have many predators in their natural environment, including snakes, monitor lizards, raptors, and a variety of other avian species [45], and they are particularly vulnerable to predation when they are fledglings [46]. Snakes and monitor lizards are especially well-documented nest predators of the species [45,47]. Therefore, we decided to present a taxidermic specimen of a lace monitor lizard, *Varanus varius*, to zebra finches.

The zebra finch has laterally positioned eyes, and in the horizontal plane, each eye has a visual field of 170° [48]. The binocular field is 30° to 40° in front and the optical axis and fovea are at 62° from the sagittal axis of the head [48]. Hence, acute vision, especially of moving stimuli, requires monocular vision, whereas the binocular field is myopic. Even grain is viewed with the lateral monocular field before the bird pecks [48]. Potential predators are also viewed in the lateral, monocular field of vision.

Therefore, our aim was to assess eye use and eye preference to view a model predator and so determine whether or not the zebra finch shows the same left eye (right hemisphere) preference to view this stimulus as found in other vertebrate species.

2. Materials and Methods

2.1. Subjects

Twenty adult zebra finches were purchased from a breeder, who housed the birds in outdoor aviaries in rural NSW, Australia. They had been exposed to species occurring naturally in this environment. These species included predators (monitor lizards and raptors), rodents, and other free-ranging animals (i.e., dogs, cats, and other native birds). At the University of New England, the zebra finches were housed in same-sex groups, in four aviaries (1.5 m × 1 m × 2 m) located in a single room.

Each aviary was furnished with branches for perching and nest baskets. Ambient temperature was maintained within a range of 18–27 °C. The light cycle was 13L:11D, the main lights were turned on at 06.00 h and off at 19.00 h. A small lamp with a 40 W globe was placed in the centre of the room and was switched on at 19.00 h and switched off at 19.30 h. The latter was to provide a cue to the birds to begin to roost and to provide some light for a brief period once the main light source had been switched off. A Hitachi 40 W fluorescent light with an UV output of 7.5 was activated for 30 min from 07.30 h to 08.00 h daily to provide a source of UV light. Food, water, and cuttlefish bone were supplied *ad libitum*. The seed used was a mixture of two commercial brands of food for finches (Lovitt and Trill). Vitamin and calcium supplements were provided once every 2 weeks and lettuce was provided once a week. The cage floor was lined with newspaper and replaced once a week.

Individual birds were identified using the following features: Female zebra finches were identified by noting their colour morph (wild type, fawn, or white morph), beak colour and size, and shape of the markings on their heads/faces. Males were first categorised by their colour morph and then the characteristics that were used to identify individual males of the same morph were beak colour, size, and shape of chest band, presence or absence of a white patch of feathers under beak, and the colour or the pattern underneath the wings. All of these markings were individually distinctive, and since the birds were housed in small home groups, identifying individuals was accurate. Ring bands were not used to identify individual birds since this could have influenced their behaviour [49].

Housing and testing of the zebra finches was conducted with the authority of the Animal Ethics Committee of the University of New England (AEC numbers 06/091, 07/014, 06/090).

2.2. Testing Room and Aviary

Testing occurred in the birds' second year of life after some pilot tests had been conducted (see below) and 9 months after the birds had been purchased and housed at the university. All experiments were conducted in an aviary (3 m × 1.5 m × 2 m) located in a room separated from where the home aviaries were kept (Figure 1). The testing aviary was divided into two virtual sections of equal size (1.5 m × 1.5 m × 2 m). Section A was the half of the aviary where the stimuli were presented, and Section B was the other half of the aviary. Two perches were placed in Section B, in the corners of the aviary furthest from location of the stimulus, from one of which the bird could see into Section A of the aviary and could see the stimulus at a distance. The other perch in Section B was located behind a visual barrier to provide a place of refuge at some distance from the stimulus.

Section A contained some branches spanning from the floor to the roof of the aviary and located at the border of Section A and B. The main perch was 80 cm in length, 40 cm from the shorter side of the aviary where the stimulus was presented and at 30 cm above a platform on which the stimulus was presented (i.e., the bird was approximately 50 cm from the platform and stimulus). The platform was 40 cm × 60 cm, located midway in the shorter side of the cage and 140 cm from the floor of the aviary.

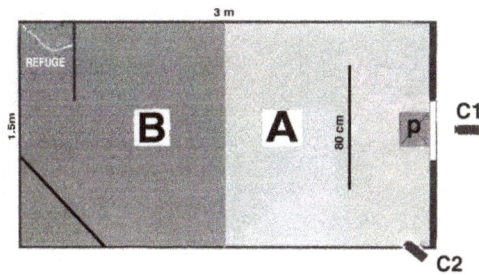

Figure 1. A view of the testing cage from above. The cage was divided into two virtual sections, **A** and **B**. The video recorders are **C1** and **C2**. **P** indicates the platform on which the stimuli were presented. See text for details. Note this diagram is not drawn to a precise scale.

Two digital video recorders (Panasonic NVGS35) were located outside the aviary (camera 1 at 120 cm from it) behind the stimulus presentation area. One video recorder was placed 30cm above the platform so that it recorded a clear, close-up view of the bird on the main perch near the stimulus. The other was placed so that a wide-range view incorporated most of the aviary with the exception of the refuge area. The experimenter was located behind a visual barrier and could observe the finches via a monitor.

2.3. Testing

Some pilot tests were conducted using birds tested alone but this was found to be unsuccessful because separation from the group caused stress. The isolated birds tended to freeze and not move or feed in the testing aviary and this was considered undesirable for the bird, and counterproductive for the testing procedure. Therefore, pairs were tested (each pair treated as N = 1) in order to avoid confounding our results with effects of stress through social isolation [50]. During these preliminary tests, the birds were between 6 and 12 months of age.

In all cases, the pairs were same-sex cage mates. Zebra finches in captivity may form same-sex pair bonds [51]. To determine whether individuals had formed a bond, each cage group was observed for 30 min per day over four consecutive days and the interactions between the individuals were noted. Seven bonded pairs (five same-sex male pairs and two same-sex female pairs) were identified.

Individuals that had formed a bond were seen allopreening and no agonistic behaviour between the pair was recorded. By contrast, agonistic events were regularly directed toward individuals not part of the pair. Only those that had formed pair-bonds were selected for testing

Collecting the finches for transport to the testing room and testing aviary was achieved by turning off the lights. Since zebra finches have poor eyesight in dim light, they could be collected from their perches without undue stress (no flight response). The experimenter was able to take each finch and place it in a small transport box and then release it into the testing aviary. The pair then remained in the test-cage for 5 days before the actual tests were conducted. This allowed them to adjust to the new surroundings before the stimuli were presented.

To present the stimuli, the experimenter placed the stimulus in the aviary through a door next to the presentation platform. The stimulus was retrieved using the same procedure. Note, only the arm of the experimenter was visible to the birds briefly while the stimulus was placed on the platform or retrieved from it. The experimenter was otherwise visually isolated from the birds throughout testing. Behaviour was scored from the video-footage.

2.4. Stimuli and Their Presentation

The non-threatening stimulus was a white opaque plastic container, hereafter referred to as a jar, with a red lid (height 16 cm, base diameter 9 cm, lid diameter 5 cm; Figure 2A). The zebra finches had been previously exposed to similar containers although not this particular one.

Figure 2. The stimuli presented to the zebra finches. (**A**) Plastic jar. (**B**) Taxidermic specimen of a monitor lizard sized relative to the jar—bar length 5 cm. (**C**) The monitor lizard enlarged to show detail and photographed from the same angle as it would be seen by the bird looking down from the perch in Section A of the cage.

The predator used, as already indicated, was a taxidermic specimen of a monitor lizard, *Varanus varanus*, 125 cm in length (Figure 2B,C).

Total time for a test was 15 min. At the end of a 5-min pre-test period, and when both the birds were in Section B of the cage, the experimenter opened a small door cut into the aviary wire and placed the stimulus on the platform. The test period was of 5-min duration, after which the stimulus was removed and then there was a post-test period of 5 min. Behaviour performed in all three periods was recorded on videotape.

Each stimulus was presented to each pair once per day at 9.30 a.m. or 10.30 a.m. for a total of 6 days. The order of presentation was random.

2.5. Data Collection

For a zebra finch to fixate a stimulus monocularly, it needs to make an exaggerated head movement. This provided an opportunity to score whether there was an eye preference to view the stimulus. After approaching a stimulus (i.e., landing on the perch directly in front of the stimulus), the zebra finches performed monocular fixation movements, described by Lombardi and Curio [43,52]. This was classified as inspection behaviour. To ensure that the bird was fixating the stimulus monocularly (i.e., for 1 or more seconds) during an inspection event, the angle of the bird's beak to the stimulus had to be $90° \pm 20°$ (Figure 3).

Figure 3. An illustration of the angle of viewing the stimulus, at a right angle to the line of the beak.

Scoring used playback of video recordings which allowed frame-by-frame analysis, particularly for scoring eye use (Numbers 4 and 5 below). The following behaviour was scored from the videotapes:

1. Latency for the first bird to enter Section A (i.e., the Section in which the stimulus was presented); see Figure 1.
2. Time spent in Section A in the pre-test, test, and post-test periods. Each bird was scored individually, and the final tally was the total for the pair.
3. Number of visits into Section A.
4. Number of monocular fixation events and alternating monocular fixation bouts per pair. These were scored only when the bird was in Section A on the perch directly in front of the stimulus.
5. Eye used in each fixation event of at least one second duration.

2.6. Statistical Analysis

Data were analysed for normality and equal variances and if the assumptions for parametric tests were not met, non-parametric statistics were conducted. The non-parametric data were analysed using Friedman's test with testing day as the repeated measure. Post hoc analyses used two-sample

Wilcoxon signed ranks tests. The parametric data were analysed using *t*-tests. Note that the sample size used was the number of pairs, not the number of individuals.

3. Results

Scores were obtained for all 7 pairs over the 6 days of testing and with both stimuli.

3.1. Time Spent in Section A, Near the Stimulus

The mean latency to move into Section A, and so approach the stimulus, was 76 ± 12 s when the lizard was presented, compared to 30 ± 8 s when the jar was presented (2-tailed, paired *t*-test, $t = -2.952, p = 0.026$).

In tests involving presentation of the jar, the time spent in Section A did not vary significantly between pre-test, test, and post-test periods (Friedman's test with period as a repeated measure, $\chi^2 = 0.240, p = 0.887$; Figure 4). By contrast, time spent in Section A did vary significantly in tests in which the lizard was presented ($\chi^2 = 17.532, p = 0.001$); during presentation of the lizard, the birds spent significantly less time in Section A than they did in the pre-test period (Wilcoxon, $Z = -3.194$, $p = 0.001$) or in the post-test period ($Z = -3.210, p = 0.001$). In other words, compared to periods without presentation of a stimulus and during presentation of the jar, the birds stayed further away from the lizard by remaining in Section B. The difference between these scores during presentation of the jar versus the monitor lizard was significant (*t*-test, $t = -2.796, p = 0.031$). Despite their avoidance of the lizard, the birds made no fewer visits from Section B into Section A when the lizard was presented compared to the number of such visits in the pre-test and post-test periods (Friedman's test, $\chi^2 = 3.920$, $p = 0.141$).

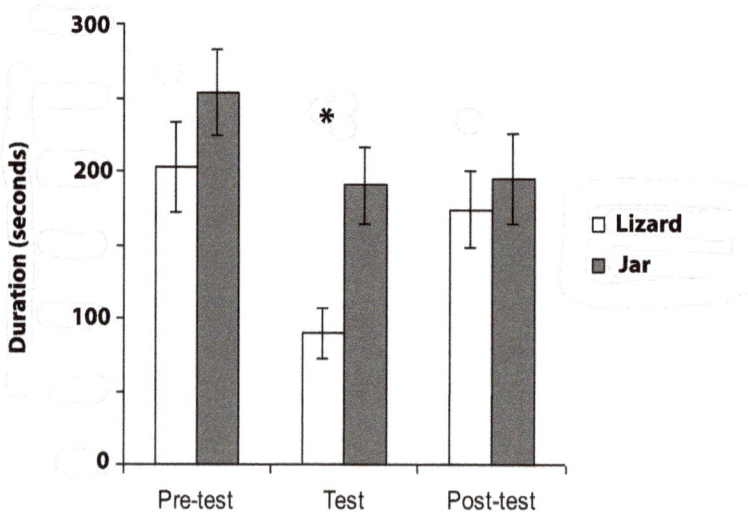

Figure 4. Time spent in Section A of the cage in the pre-test, test, and post-test period, each of 5 min. Asterisk indicates a significant difference between the time spent during presentation of the lizard versus the jar (see text for details). The birds spent less time in Section 1 when the lizard was presented, which indicates that it was perceived as threatening.

3.2. General Characteristics of a Looking Event

The zebra finches fixated the stimulus (i.e., for one or more seconds) monocularly and did so either in a single monocular viewing event or by alternating between the monocular fields of each eye, turning the head from side to side. Single monocular fixations occurred in $39 \pm 8\%$ of the viewing events when the jar was presented and $43 \pm 5\%$ when the lizard was presented (no significant difference, 2-tailed paired *t*-test $p = 0.69$).

In biocular alternating fixation bouts (looking first with monocular field of one eye and then turning the head to look with the monocular field of the other eye and so on; see Introduction and [44]), the number of monocular fixations per bout was a mean and standard error of 2.28 ± 0.35 for the jar and 2.66 ± 0.26 for the lizard (not significantly different, 2-tailed *t*-test, $p = 0.40$). Nevertheless, when viewing the jar, the birds were more likely to end a fixation bout using the eye opposite to the one with which they had begun the bout than they were to end the bout viewing with the same eye with which they had started (2-tailed paired *t*-test, $p = 0.001$). This means that, to view the jar, the alternating fixation bouts were mostly LR or RL, where L refers to left eye and R to right eye (Figure 5). When viewing the lizard, there was no difference between the number of bouts with odd versus even numbers of fixations (2-tailed paired *t*-test, $p = 0.471$; Figure 5). In other words, when viewing the lizard LRL was as common as LR (see later for eye preference).

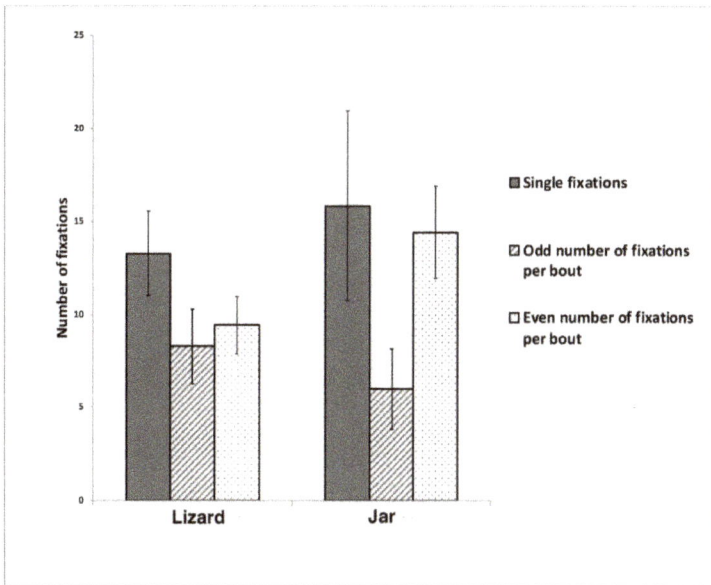

Figure 5. Types of looking bouts. The mean number of each type of looking bout (with standard error bar) is presented for all days of testing. The categories are single fixation bouts (fixating the stimulus with one eye only), biocular alternating fixation bouts with an odd number of fixations (left-right-left or right-left-right), and biocular fixation bouts with an even number of fixations (left-right or right-left). Very occasionally the alternating fixation bouts were longer than the examples stated in the brackets, and these were included in the appropriate category.

The total number of fixation events (monocular looking without alternation plus biocular alternating fixation) did not differ between the two stimuli (mean and standard error of the mean, sem, for the jar was 35.43 ± 7.27 and for the lizard was 29.71 ± 5.77; U-test comparison $p = 0.599$). Therefore, although the birds spent less time in Section A of the cage during presentations of the lizard compared to presentations of the jar, when they were in Section A of the cage, they viewed both of these stimuli to the same amount.

However, differences did occur between stimuli when comparison was made of the number of inspection events across days. The pattern of responses across days differed for the two stimuli.

As shown in Figure 6, on presentation of the jar, the events of fixation declined significantly across days (Friedman's test with testing day as a repeated measure, $\chi^2 = 12.14$, $p = 0.033$). The decrease from day 1 to day 2 did not reach significance (Wilcoxon, 1-tailed, $p = 0.061$) but the decrease from day 1 to day 3 was significant (Wilcoxon test, 1-tailed, $p = 0.028$). Habituation of the response had occurred. By comparison, no significant habituation of the number of inspection events was found in tests with presentation of the lizard ($\chi^2 = 8.233$, $p = 0.144$).

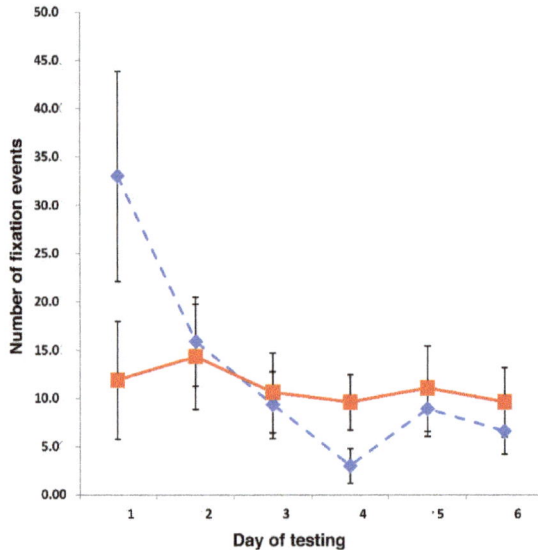

Figure 6. The mean number of fixation bouts is plotted for each stimulus and for each day of testing. Standard error bars are marked. Blue diamonds and dotted lines are for presentations of the jar. Red squares and lines are for presentations of the lizard. See text for details.

3.3. Eye Preferences

The eye used to fixate the stimulus was determined. These scores included monocular fixation of one or more seconds duration, in bouts without alternation and in bouts of alternating monocular viewing (i.e., all fixations were included regardless of whether there was a single fixation or a bout of biocular alternating fixations). Scores from all trials were included in the initial analysis. A significant left eye preference was found for viewing the lizard (percent left eye preference = $56 \pm 2\%$, 1-tailed t-test, $p = 0.001$, direction of difference predicted). This compared to no significant eye preference to view the jar ($49 \pm 3\%$, 2-tailed t-test, $p = 0.681$). The difference between eye used to view the stimuli was significant (1-tailed, paired t-test, $p = 0.031$).

Considering only the first fixation per bout (i.e., eye used to view the stimulus in a single bout or at the start of an alternating bout) calculated over all days, a significant preference to use the left eye to view the lizard was found ($65 \pm 7\%$ left eye; 1-tailed t-test, $p = 0.034$). For viewing the jar, there was no significant eye preference ($42 \pm 6\%$ left eye; 2-tailed, t-test, $p = 0.230$). The difference in eye preference to view the two stimuli was significant (2-tailed, paired t-test, $p = 0.023$).

We noticed 37 occasions on which the bird initially glanced at the stimulus for much less than a second and immediately switched to using the other eye to fixate it. When the lizard was presented, 17 of such switches were from the right eye to the left eye, compared to only two from left eye to right eye. This is consistent with the preference to view the lizard using the left eye. When the jar was presented, 18 switches were recorded, and half of these were from left eye to fixate with the right eye and half in the opposite direction (i.e., revealing no eye preference).

Next the data were analysed using only the first monocular look on the first day of presentation. The percent left eye preference was $72.4 \pm 13.5\%$ for the lizard and $42.8 \pm 12.6\%$ for the jar (Figure 7: 1-tailed *t*-test for the comparison of these two groups, $p = 0.031$). Preferred use of the left eye to view the lizard was, therefore, evident from the very first look at the lizard, compared to no significant eye preference to look at the jar.

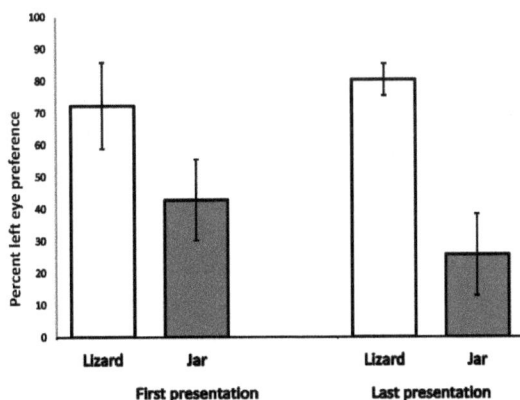

Figure 7. The mean percent left-eye preference is plotted with mean and standard error bars for the first and the last presentation of each stimulus. Note the significant left-eye preference to view the lizard, compared to no significant eye preference to view the jar (see text below).

In fact, these eye preferences remained unchanged across presentations. On the last trial, the percent left eye preference was $80.4 \pm 4.9\%$ for the lizard (significant left eye preference, 1-tailed *t*-test, $p = 0.008$) and $25.7 \pm 12.8\%$ for the jar (Figure 7: 1-tailed *t*-test for the comparison of these two groups, $p = 0.005$). Although from the first to the last presentation of the jar there was trend towards more use of the right eye (2-tailed paired *t*-test comparison of first to last presentation of the jar, $p = 0.095$), this did not reach a significant right eye preference (2-tailed, paired *t*-test, $p = 0.107$).

4. Discussion

First it was important to obtain evidence that the zebra finches perceived the lizard, and not the jar, as a threat. Therefore, approach to the stimuli was measured. Latency to move into Section A after presentation of the stimulus was one such measure. This latency was twice as long when the lizard was presented than it was when the jar was presented. Also, the birds spent significantly less time in Section A of the cage when the lizard was presented compared to the time that they spent in Section A during the pre-test and post-test periods. By contrast, no avoidance of Section A occurred when the jar was presented. Both of these measures confirmed that the zebra finches perceived the lizard, and not the jar, as a threat (i.e., recognised it as a predator).

On the first presentation of the jar, the birds fixated it more times than they did the lizard or than they did on any subsequent presentation. This suggests that, on first presentation, the jar might have been perceived as somewhat novel, but not threatening. By contrast, response to the lizard remained unchanged across the 6 days of testing; this lack of habituation of response to the lizard is likely to be caused by elevated levels of fear induced by this potential predator [53]. Despite these differences in response to the two stimuli, the total number of times the birds fixed the stimuli, over all days of testing, were the same for the lizard and the jar. However, use of the eyes to view the lizard compared to the jar did differ. There was a significant preference to view the lizard using the left eye but no significant eye preference to view the jar. This finding of a left eye preference to view the lizard was reinforced by occasions when the birds caught sight of this stimulus with the right eye and immediately switched to fixate it with the left eye. Although switches were also observed on presentation of the jar, there was no directional preference to switch from right to left eye versus from left to right eye to fixate the stimulus.

Lombardi and Curio [43,52] described the zebra finch's response to seeing a live owl as including side to side movements of the head which allow monocular fixation. This sequential use of the left and right monocular visual fields was also observed in our study but single monocular fixations with one eye were also common. When the stimuli were viewed using monocular alternating fixation bouts, these differed for viewing the lizard versus the jar. Such bouts to view the jar most often involved a fixation with one eye and then the other, whereas for viewing the lizard the birds often looked with one eye (left eye preference), then the other eye, and again with the first eye (left eye again).

In fact, the finding of a left eye preference to view the lizard versus no preferred eye to view the jar remained consistent across all days of testing. A preference to use the left eye to view a predator has also been reported in other avian species. Australian magpies, *Gymnorhina tibicen*, display a left eye preference to view their main predator, a wedge-tailed eagle, prior to withdrawing from it [39,40,54], and also in our tests the zebra finches moved away from the monitor lizard after they had inspected it from the vantage point of the perch in Section A of the cage. Domestic chicks show a left-eye preference to monitor a model of a raptor, and this contrasts with use of the right eye to search for grains of food [30,55]. Domestic chicks also show a left eye preference to monitor biological motion [56], a finding that fits well with use of the left eye to view predators.

Such left eye preference for detecting and viewing predators extends to other vertebrate species. A Dasyurid marsupial has been shown to respond more strongly to a predator, a model snake, seen in its left lateral field of vision compared to the same stimulus seen its right lateral field of vision [42], and the same has been found in the cane toad, *Bufo marinus* [41]. Therefore, together with our current finding in the zebra finch, the evidence for left eye preference to view a predator before responding by withdrawing from it is consistent across a range of species. Since inputs from the left eye are processed almost entirely by the right hemisphere, left-eye preference reflects specialisation of the right hemisphere for processing of and retreating from predators (summarised in [32]).

Butler et al. [44] also found a left eye preference in starlings viewing model predators (hawks), but they used a different method—a number of fixations of the stimulus made with one eye (using different regions of the retina) before switching to view the stimulus with the other eye. The left eye preference measured in this way was not specific to viewing predators since it was also found when the stimulus was simply a patch of grass similar to one on which one of the predators had been presented. The researchers suggested that the left eye preference resulted from a higher concentration of single cones in the left eye compared to the right eye [57]. However, we think this is unlikely to be the sole reason for the left eye preference, since by intracranial injections into the left or right hemisphere of the chick, it has been shown that asymmetry of eye use for visual discrimination learning depends on hemispheric differences in processing stimuli and controlling response (viz., right eye and left hemisphere for visual discrimination learning [22], summarised in [58]). Furthermore, if the response to a predator is close approach with contact during attack rather than withdrawal, as is often the case in magpies, there is a preference for use of the right eye and left hemisphere just before attacking [40].

Considerable research of the regions of the zebra finch brain receiving visual inputs has been undertaken. As mentioned in the Introduction, most of the visual input from an eye is processed by the contralateral hemisphere. In fact, the first relay station for visual inputs in the main visual pathway is entirely to the optic tectum contralateral to the eye. By far the majority of neurones from each optic tectum projects to the ipsilateral nucleus rotundus, although a minority do cross the midline to project to the contralateral nucleus rotundus [59]. Electrophysiological recordings of neurones in the nucleus rotundus have shown that, while there are neurones that respond to inputs from both the ipsilateral and contralateral eye, there are no neurones that respond exclusively to the ipsilateral eye [60]. Consequently, Schmidt and Bischof [60] suggested that at rotundal level, there is inhibition of input from the ipsilateral eye by contralateral input and thereby only one eye engages the bird's attention.

From each nucleus rotundus projections go to the forebrain only on the same side. Therefore, inputs from one eye are processed on the contralateral side of the forebrain (in the contralateral hemisphere). While one eye is attending to a particular stimulus, visual processing of inputs by the other eye is suppressed, as shown by Voss and Bischof [61]. In fact, the suppressed eye moves in a saccade counter to the attending eye so that foveal inputs from only the attending eye are processed at forebrain level [62]. Hence, we may deduce that attention to a predator using the left eye engages the right hemisphere and suppresses information coming from the right eye.

Taking all of the available evidence of asymmetry in zebra finches into account indicates that lateralised brain function is the same as in other avian species. At the least, use of the right eye and left hemisphere to discriminate grain from pebbles [34] and the left eye and right hemisphere to attend to predators is consistent with the pattern of lateralisation in chicks, magpies [63], and other avian species (see Introduction). One apparent discrepancy remains, and that concerns copulation and courtship. Whereas male zebra finches use their right eye to approach a female in courtship and their songs have more motifs when they see the female by using their right eye (discussed in the Introduction), the evidence from the domestic chick is that the left eye and right hemisphere control the copulation response [64]. It is possible that the difference depends on courtship behaviour versus actual performance of copulation. During courtship performance, copulation behaviour must be suppressed and that could be achieved by the left hemisphere's ability to suppress the right hemisphere [31], and hence the right eye/left hemisphere is used during courtship. Consistent with this explanation, in sage-grouse Krakauer et al. [65] found significant left-eye preference during courtship only in males that mated successfully and not in those that were unsuccessful in mating.

5. Conclusions

As we outlined in the Introduction, despite early indications that the direction of asymmetry in the zebra finch brain concerning song production and perception differed from that of other songbird species, more recent evidence indicates that any such difference may be minimal, although further research is needed to confirm this suggestion. However, taking into account the evidence for asymmetry of visual behaviour, the pattern of asymmetry in this species seems to match that of other avian species—specialisation of the left hemisphere for focussed attention to perform routine functions in non-stressful situations, and of the right hemisphere for broad attention, response to novel stimuli, and control of behaviour in emergency (predatory) situations (summarised in [31,32]).

Author Contributions: G.K. and A.K. conceived and designed the experiments; author G.K. supervised the project; author A.K. performed the experiments and cared for the birds; authors L.J.R. and A.K. analysed the data; L.J.R. wrote the paper.

Funding: The project was funded by The Cardigan Fund, an annual bequest to G.K. A.K. received a postgraduate scholarship (Ph.D.) from the University of New England.

Conflicts of Interest: The authors declare no conflict of interest. There were no funding sponsors that had any role in the writing of this manuscript or in any other capacity in preparing and publishing this manuscript.

References

1. Van Ruijssevelt, L.; Washington, S.D.; Hamaide, J.; Verhoye, M.; Keliris, G.A.; Van der Linden, A. Song processing in the zebra finch auditory forebrain reflects asymmetric sensitivity to temporal and spectral structure. *Front. Neurosci.* **2017**, *11*, 549. [CrossRef] [PubMed]
2. Immelman, K. Song Development in the zebra finch and other estrildid finches. In *Bird Vocalizations*; Hinde, R.A., Ed.; Cambridge University Press: Cambridge, UK, 1969; pp. 61–77.
3. Slater, P.J.B.; Richards, C.; Mann, N.I. Song learning in zebra finches exposed to a series of tutors during the sensitive phase. *Ethology* **1991**, *88*, 163171. [CrossRef]
4. Doupe, A.J.; Kuhl, P.K. Birdsong and human speech: Common themes and mechanisms. In *Neuroscience of Birdsong*; Zeigler, H.P., Marler, P., Eds.; Cambridge University Press: Cambridge, UK, 2008; pp. 5–31.

5. Benichov, J.I.; Benezra, S.; Vallentin, D.; Globerson, E.; Long, M.; Tchemichovski, O. The forebrain song system mediates predictive call timing in female and male zebra finches. *Curr. Biol.* **2016**, *26*, 309–318. [CrossRef] [PubMed]

6. Olson, E.M.; Maeda, R.K.; Gobes, S.M.H. Mirrored patterns of lateralized neuronal activation reflect old and new memories in the avian auditory cortex. *Neuroscience* **2016**, *330*, 395–402. [CrossRef] [PubMed]

7. Yang, L.M.; Vicario, D.S. Exposure to a novel stimulus environment alters patterns of lateralization in avian auditory cortex. *Neuroscience* **2015**, *285*, 107–118. [CrossRef] [PubMed]

8. Nottebohm, F. Neural lateralization of vocal control in a passerine bird. I. Song. *J. Exp. Zool.* **1971**, *177*, 229–261. [CrossRef] [PubMed]

9. Nottebohm, F. Asymmetries in neural control of vocalization in the canary. In *Lateralization in the Nervous System*; Harnard, S., Doty, R.W., Goldstein, L., Jaynes, J., Krauthamer, G., Eds.; Academic Press: New York, NY, USA, 1977; pp. 23–44.

10. Nottebohm, F.; Alvarez-Buylla, A.; Cynx, J.; Kirn, J.; Ling, C.Y.; Nottebohm, M. Song learning in birds: The relation between perception and production. *Philos. Trans. R. Soc. Lond. B* **1990**, 115–124. [CrossRef] [PubMed]

11. Nottebohm, F.; Arnold, A.P. Sexual dimorphism in vocal control areas of the songbird brain. *Science* **1976**, *194*, 211–213. [CrossRef] [PubMed]

12. Williams, H.; Crane, L.A.; Hale, T.K.; Esposito, M.A.; Nottebohm, F. Right-side dominance for song control in the zebra finch. *J. Neurobiol.* **1992**, *23*, 1006–1020. [CrossRef] [PubMed]

13. Cynx, J.; Williams, H.; Nottebohm, F. Hemispheric differences in avian song discrimination. *Proc. Natl. Acad. Sci. USA* **1992**, *89*, 1372–1375. [CrossRef] [PubMed]

14. Floody, O.R.; Arnold, A.P. Song lateralization in the zebra finch. *Horm. Behav.* **1997**, *31*, 25–34. [CrossRef] [PubMed]

15. Voss, H.U.; Tabelow, K.; Polzehl, J.; Tchernichovski, O.; Maul, K.K.; Salgado-Commissariat, D.; Ballon, D.; Helekar, S.A. Functional MRI of the zebra finch brain during song stimulation suggests a lateralized response topography. *Proc. Natl. Acad. Sci. USA* **2007**, *104*, 10667–10672. [CrossRef] [PubMed]

16. Poirier, C.; Bourmans, T.; Verhoye, M.; Balthazart, J.; Van der Linden, A. Own song recognition in the songbird auditory pathway: Selectivity and lateralization. *J. Neurosci.* **2009**, *29*, 2252–2258. [CrossRef] [PubMed]

17. Avery, M.T.; Phillmore, L.S.; MacDougall-Shackleton, S.A. Immediate early gene expression following exposure to acoustic and visual components of courtship in zebra finches. *Behav. Brain Res.* **2006**, *165*, 247–253.

18. Moorman, S.; Gobes, S.M.H.; Kuijpers, M.; Kerkofs, A.; Zandbergen, M.A.; Bolhuis, J.J. Human-like brain hemispheric dominance in birdsong learning. *Proc. Natl. Acad. Sci. USA* **2012**, *109*, 12782–12787. [CrossRef] [PubMed]

19. Lampen, J.; McAuley, J.D.; Chang, S.-E.; Wade, J. ZENK induction in the zebra finch brain by song: Relationship to hemisphere, rhythm, oestradiol and sex. *J. Neuroendocrin.* **2017**, *29*, e12543. [CrossRef] [PubMed]

20. Van Ruijssevelt, L.; Chen, Y.; von Eugen, K.; Hamaide, J.; De Groof, G.; Verhoye, M.; Güntürkün, O.; Woolley, S.C.; Van der Linden, A. fMRI reveals a novel region for evaluating acoustic information for mate choice in a female songbird. *Curr. Biol.* **2018**, *28*, 711–721. [CrossRef] [PubMed]

21. Slater, P.J.B.; Eales, L.A.; Clayton, N.S. Song learning in zebra finches (*Taeniopygia guttata*): Progress and prospects. *Adv. Study Behav.* **1988**, *18*, 1–34.

22. Rogers, L.J.; Anson, J.M. Lateralisation of function in the chicken fore-brain. *Pharm. Biochem. Behav.* **1979**, *10*, 679–686. [CrossRef]

23. Workman, L.; Andrew, R.J. Asymmetries of eye use in birds. *Anim. Behav.* **1986**, *34*, 1582–1584. [CrossRef]

24. Ten Cate, C.; Baauw, A.; Ballintijn, M.; van der Horst, I. Lateralization of orientation in sexually active zebra finches: Eye use asymmetry of locomotion bias? *Anim. Behav.* **1990**, *39*, 992–994. [CrossRef]

25. Workman, L.; Andrew, R.J. Population lateralization in zebra finch courtship: An unresolved issue. *Anim. Behav.* **1991**, *41*, 545–546. [CrossRef]

26. Ten Cate, C. Population lateralization in zebra finch courtship: A re-assessment. *Anim. Behav.* **1991**, *41*, 900–901. [CrossRef]

27. George, I.; Hara, E.; Hessler, N.A. Behavioral and neural lateralization of vision in courtship singing of the zebra finch. *J. Neurobiol.* **2006**, *66*, 1164–1173. [CrossRef] [PubMed]
28. Weidner, C.; Reperant, J.; Miceli, D.; Haby, M.; Rio, J.P. An anatomical study of ipsilateral retinal projections in the quail using autoradiographic, horseradish peroxidase, fluorescence and degeneration technique. *Brain Res.* **1985**, *340*, 99–108. [CrossRef]
29. Lieshoff, C.; Grosse-Ophoff, J.; Bischof, H.-J. Sexual imprinting leads to lateralized and non-lateralized expression of the immediate early gene zenk in the zebra finch. *Behav. Brain Res.* **2004**, *148*, 145–155. [CrossRef]
30. Rogers, L.J. Evolution of hemispheric specialisation: Advantages and disadvantages. *Brain Lang.* **2000**, *73*, 236–253. [CrossRef] [PubMed]
31. Rogers, L.J.; Vallortigara, G.; Andrew, R.J. *Divided Brains: The Biology and Behaviour of Brain Asymmetries*; Cambridge University Press: Cambridge, UK, 2013.
32. MacNeilage, P.; Rogers, L.J.; Vallortigara, G. Origins of the left and right brain. *Sci. Am.* **2009**, *301*, 60–67. [CrossRef] [PubMed]
33. Templeton, J.J.; McCracken, B.G.; Sher, M.; Mountjoy, D.J. An eye for beauty: Lateralized visual stimulation of courtship behaviour and mate preferences in male zebra finches, *Taeniopygia guttata*. *Behav. Proc.* **2014**, *102*, 33–39. [CrossRef] [PubMed]
34. Alonso, Y. Lateralization of visual guided behavior during feeding in zebra finches (*Taeniopygia guttata*). *Etología* **1997**, *5*, 67–72.
35. Alonso, Y. Lateralization of visual guided behavior during feeding in zebra finches (*Taeniopygia guttata*). *Behav. Proc.* **1998**, *43*, 257–263. [CrossRef]
36. Mench, J.; Andrew, R.J. Lateralisation of a food search task in the domestic chick. *Behav. Neural Biol.* **1986**, *46*, 107–114. [CrossRef]
37. Güntürkün, O.; Kesch, S. Visual lateralization during feeding in pigeons. *Behav. Neurosci.* **1987**, *101*, 433–435. [CrossRef] [PubMed]
38. Güntürkün, O. The ontogeny of visual lateralization in pigeons. *Ger. J. Psychol.* **1993**, *17*, 276–287.
39. Rogers, L.J.; Kaplan, G. An eye for a predator: Lateralization in birds, with particular reference to the Australian magpie. In *Behavioral and Morphological Asymmetries in Vertebrates*; Malashichev, Y., Deckel, W., Eds.; Landes Bioscience: Austin, TX, USA, 2006; pp. 47–57.
40. Koboroff, A.; Kaplan, G.; Rogers, L.J. Hemispheric specialization in Australian magpies (*Gymnorhina. tibicen*) shown as eye preferences during response to a predator. *Brain Res. Bull.* **2008**, *76*, 304–306. [CrossRef] [PubMed]
41. Lippolis, G.; Bisazza, A.; Rogers, L.J.; Vallortigara, G. Lateralization of predator avoidance responses in three species of toads. *Laterality* **2002**, *7*, 163–183. [CrossRef] [PubMed]
42. Lippolis, G.; Westman, W.; McAllan, B.M.; Rogers, L.J. Lateralization of escape responses in the striped-faced dunnart, *Sminthopsis macroura* (Dasyuridae: Marsupalia). *Laterality* **2005**, *10*, 457–470. [CrossRef] [PubMed]
43. Lombardi, C.M.; Curio, E. Social facilitation of mobbing in the zebra finch *Taeniopygia guttata*. *Bird Behav.* **1985**, *6*, 34–40. [CrossRef]
44. Butler, S.R.; Templeton, J.J.; Fernández-Juricic, E. How do birds look at their world? A novel avian visual fixation strategy. *Behav. Ecol. Sociobiol.* **2018**, *72*, 38. [CrossRef]
45. Zann, R. *The Zebra Finch: A Synthesis of Field and Laboratory Studies*; Oxford University Press: New York, NY, USA, 1996.
46. Zann, R.; Runciman, D. Survivorship, dispersal and sex ratios of zebra finches *Taeniopygia guttata* in southeastern Australia. *Ibis* **1994**, *136*, 136–146. [CrossRef]
47. Immelmann, K. Beiträge zu einer vergleichenden Biologie australische Prachtfinken (Spermestidae). *Zoologische Jahrbücher Abteilung für Systematik Okologie und Geograohie der Tiere* **1962**, *90*, 1–196.
48. Bischof, H.-J. The visual field and visually guided behaviour in the zebra finch (*Taenopygia guttata*). *J. Comp. Physiol. A* **1988**, *163*, 329–337. [CrossRef]
49. Burley, N. Comparison of band colour preferences in two species of estrildid finches. *Anim. Behav.* **1986**, *34*, 1732–1741. [CrossRef]
50. Mainwaring, M.C.; Beal, J.L.; Hartley, I.R. Zebra finches are bolder in an asocial, rather than a social, context. *Behav. Proc.* **2011**, *87*, 171–175. [CrossRef] [PubMed]

51. Adkins-Regan, E.; Krakauer, A. Removal of adult males from the rearing environment increases preference for same-sex partners in the zebra finch. *Anim. Behav.* **2000**, *60*, 47–53. [CrossRef] [PubMed]
52. Lombardi, C.M.; Curio, E. Influence of environment on mobbing by zebra finches. *Bird Behav.* **1985**, *6*, 28–33. [CrossRef]
53. Wallace, K.J.; Rosen, J.B. Predator odor as an unconditioned fear stimulus in rats: Elicitation of freezing by trimethylthiazoline, a component of fox feces. *Behav. Neurosci.* **2000**, *114*, 912–922. [CrossRef] [PubMed]
54. Kaplan, G.; Rogers, L.J. Stability of referential signalling across time and locations: Testing calls of Australian magpies (*Gymnorhina tibicen*) in urban and rural Australia and Fifi. *PeerJ* **2013**, *1*, e112. [CrossRef] [PubMed]
55. Rogers, L.J.; Zucca, P.; Vallortigara, G. Advantage of having a lateralized brain. *Proc. Roy. Soc. Lond. B* **2004**, *271*, S420–S422. [CrossRef] [PubMed]
56. Rugani, R.; Rosa Salva, O.; Regolin, L.; Vallortigara, G. Brain asymmetry modulates perception of biological motion in newborn chicks (*Gallus gallus*). *Behav. Brain Res.* **2015**, *290*, 1–7. [CrossRef] [PubMed]
57. Hart, N.S.; Partridge, J.C.; Cuthill, I.C. Retinal asymmetry in birds. *Curr. Biol.* **2000**, *10*, 115–117. [CrossRef]
58. McCabe, B. Pharmacological agents and electrophysiological techniques. In *Lateralized Brain Functions*; Rogers, L.J., Vallortigara, G., Eds.; Humana Press: New York, NY, USA, 2017; pp. 251–276.
59. Hermann, K.; Bischof, H.-J. Development of the tectofugal visual system of normal and deprived zebra finches. In *Vision, Brain, and Behavior in Birds*; Zeigler, H.P., Bischof, H.-J., Eds.; The MIT Press: Cambridge, MA, USA, 1993; pp. 207–226.
60. Schmidt, A.; Bischof, H.-J. Integration of information from both eyes by single neurons of neucleus rotundus, ectostriatum and lateral neostriatum of the zebra finch (*Taeniopygia guttata castanotis* Gould). *Brain Res.* **2001**, *923*, 20–31. [CrossRef]
61. Voss, J.; Bischof, H.-J. Regulation of ipsilateral visual information within the tectofugal visual system in zebra finches. *J. Comp. Physiol. A* **2003**, *189*, 545–553. [CrossRef] [PubMed]
62. Voss, J.; Bischof, H.-J. Eye movements of laterally eyed birds are not independent. *J. Exp. Biol.* **2009**, *212*, 1568–1575. [CrossRef] [PubMed]
63. Kaplan, G. Audition and hemispheric specialization in songbirds and new evidence from Australian magpies. *Symmetry* **2017**, *9*, 99. [CrossRef]
64. Rogers, L.J.; Zappia, J.V.; Bullock, S.P. Testosterone and eye-brain asymmetry for copulation in chickens. *Experientia* **1985**, *41*, 1447–1449. [CrossRef]
65. Krakauer, A.H.; Blundell, M.A.; Scanlan, T.N.; Wechsler, M.S.; McCloskey, E.A.; Yu, J.H.; Patricelli, G.L. Successfully mating male sage-grouse show greater laterality in courtship and aggressive interactions. *Anim. Behav.* **2016**, *111*, 261–267. [CrossRef]

symmetry

MDPI

Article

A Crucial Role of Attention in Lateralisation of Sound Processing?

Martine Hausberger [1], Hugo Cousillas [1], Anaïke Meter [2], Genta Karino [3], Isabelle George [1], Alban Lemasson [2] and Catherine Blois-Heulin [2,*]

[1] C.N.R.S., Ethologie animale et humaine EthoS, UMR 6552, Université de Rennes, Université Caen Normandie, Campus de Beaulieu, B25, Avenue du Général Leclerc, 35000 Rennes, France; martine.hausberger@univ-rennes1.fr (M.H.); hugo.cousillas@univ-rennes1.fr (H.C.); isabelle.george@univ-rennes1.fr (I.G.)
[2] C.N.R.S., Ethologie animale et humaine EthoS, UMR 6552, Université de Rennes, Université Caen Normandie, Station Biologique, 35380 Paimpont, France; annick.meter@univ-rennes1.fr (A.M.); alban.lemasson@univ-rennes1.fr (A.L.)
[3] Department of Biotechnology and Life Science, Graduate School of Engineering, Tokyo University of Agriculture and Technology, 2-24-16 Naka-cho, Koganei-shi, Tokyo 184-8588, Japan; gentachan2011@gmail.com
* Correspondence: catherine.blois-heulin@univ-rennes1.fr; Tel.: +33-299-61-81-65, Fax: +33-299-61-81-88

Received: 22 November 2018; Accepted: 24 December 2018; Published: 3 January 2019

Abstract: Studies on auditory laterality have revealed asymmetries for processing, particularly species-specific signals, in vertebrates and that each hemisphere may process different features according to their functional "value". Processing of novel, intense emotion-inducing or finer individual features may require attention and we hypothesised that the "functional pertinence" of the stimuli may be modulating attentional processes and hence lateralisation of sound processing. Behavioural measures in "(food) distracted" captive Campbell's monkeys and electrophysiological recordings in anesthetised (versus awake) European starlings were performed during the broadcast of auditory stimuli with different functional "saliences" (e.g., familiar/novel). In Campbell's monkeys, only novel sounds elicited lateralised responses, with a right hemisphere preference. Unfamiliar sounds elicited more head movements, reflecting enhanced attention, whereas familiar (usual in the home environment) sounds elicited few responses, and thus might not be arousing enough to stimulate attention. In starlings, in field L, when awake, individual identity was processed more in the right hemisphere, whereas, when anaesthetised, the left hemisphere was more involved in processing potentially socially meaningless sounds. These results suggest that the attention-getting property of stimuli may be an adapted concept for explaining hemispheric auditory specialisation. An attention-based model may reconcile the different existing hypotheses of a Right Hemisphere-arousal/intensity or individual based lateralisation.

Keywords: hemispheric specialisation; attention; starlings; Campbell's monkeys; auditory perception

1. Introduction

At the time of and also because of Broca's (1861) [1] early findings of a dominance of the left hemisphere for language production and processing, brain lateralisation has long been considered a unique human feature. Only in the last decades have parallels been sought and found in animals, revealing that brain lateralisation is a rather universal feature amongst vertebrates and some invertebrates [2–5]. Surprisingly, auditory laterality is amongst the latest studied aspects, but these studies have shown that there are clear asymmetries for processing, in particular species-specific sound signals, in vertebrates [6–9]. Most of these studies have investigated whether animals, as a parallel to language processing, had a dominant hemisphere for the processing of the species-specific vocalisations. Indeed, a left dominance for species-specific vocalisations has been found in a series of species such as seals [10], mice [11], raptors [12], cats [13], rhesus macaques [14,15], and chimpanzees [16] using ear orientation in response to playbacks or lesional approaches such as Heffner and Heffner [17,18]. However, results in songbirds and some primate species are more mitigated: lesions, electrophysiological and/or behavioural tests reveal a left dominance in Bengalese finches [19], rhesus macaques [20] but a right dominance in zebra finches [21], European starlings [22,23], vervet monkeys [24] and Japanese macaques [25] in response to species-specific vocal signals. When investigating further, however, both Cynx et al. [21] and George [22,23] found a more complex process as each hemisphere seemed to process different features even within the species-specific songs according to their social (e.g., individual) or functional (familiar/nonfamiliar) "value".

This reminds one of the processing of other important features of speech such as prosody and emotional content that are processed in the right hemisphere by humans [26]. These features are important for a listener to appreciate the emitter's identity, intentions and attitudes [27].

It has been proposed that, in birds, the right hemisphere would be more involved in finer discriminations [21] or responses to novel features [28], a parallel with baboons or gray mouse lemurs where it has been suggested that non-familiar sounds are processed in this hemisphere [29,30] although recent findings show that familiar stimuli are processed more in this hemisphere in Japanese macaques [25]. In Campbell's monkeys confronted by species-specific and heterospecific social calls with different emotional valences, only the species-specific calls with a negative valence elicited a lateralised response with a preference for the right hemisphere (left head turning) [31], as also observed in Emei music frogs [32]. Sex differences may occur as in mouse lemurs, for example, a species where males, but not females, exhibit a significant right ear-left hemisphere bias when exposed to conspecific communication sounds of negative emotional valence [33]. Interestingly, these laterality biases may extend to interspecific perception: in dogs, the right hemisphere dominance for conspecific barks extends to the signals of another (familiar) species, cats, while human orders are processed without any hemispheric preference [34]; in cats, the left hemisphere is more involved in the processing of species-typical vocalisations such as meow or purring, but not for growling, while sounds eliciting intense emotions (dogs' vocalisations of "disturbance") are associated with the right hemisphere [13].

In any case, finer discriminations and processing of novel or intense emotion-inducing features may require more attention, which is considered as one basis for the evolution of lateralisation [35]. Female free-ranging orcas, but not males, show a significant preference for the use of the left eye when looking at humans, which can be associated with their higher sustained visual attention towards humans [36]. It was proposed that the two hemispheres did not have a similar function: focused attention will be processed by left hemisphere and conversely broad attention by right hemisphere [37].

While most studies involve behavioural responses (i.e., eye or ear/head turning) as indirect information of brain lateralised processing, some electrophysiological studies suggest further a link between perceptual laterality and attention. Thus, in European starlings, an auditory dominance can only be observed in awake birds and not in anaesthetised birds, but also the types of sounds processed by the two hemispheres differ between the two states [22,23,38].

More recently, an EEG study on horses has revealed that processing of visual attention per se is lateralised, with a clear predominant involvement of the right hemisphere [39].

The aim of the present study was therefore to investigate this mutual relationship between attention and lateralisation, by looking at auditory perception in two species of primates and songbirds, the Campbell's monkeys and the European starlings, both known to perceive and process species-specific stimuli at least with lateralised responses (e.g., [22,23,31,40]). On the other hand, both species show different levels of attention and auditory response according to the social familiarity or social functional significance of the stimuli: in Campbell's monkeys, old (no longer used) but familiar variants of contact calls [41] or unexpected types of vocal interactions [42] elicit a cessation of activity and visual attention. Female starlings respond to the playback of familiar shared songs with visual search [43], and familiarity was a major modulation factor in auditory responses in the field L of adult male starlings [44].

We hypothesised therefore that the "functional pertinence" of the stimuli may be modulating attentional processes (see also the "attention neurons" proposed by Hubel et al. [45]) and hence lateralisation of the sound processing, while attentional processes per se would modulate responses. While we expect more functionally salient stimuli (e.g., social calls) to be triggering more attention and enhanced lateralisation, manipulating attention should change those responses. To test this hypothesis, we used behavioural tests in Campbell's monkeys and electrophysiological recordings in European starlings to investigate the impact of attention on the lateralisation of neuronal or behavioural responses to the broadcast of auditory stimuli with different functional "saliences" (familiar/novel, species-specific/non-specific, etc.). In both species, only females were tested. In European starlings, data on male lateralisation of sound processing were already known [22,23,46] and could be used for comparison. In Campbell's monkeys, females are at the core of the social network, with clear individualised bonds [47,48], and therefore appeared as interesting to test. We manipulated the attention of the animals in two ways: by adding a (food) distractor to the monkeys, as we expected them to then pay attention only to particularly salient stimuli and by looking at neuronal responses in the primary auditory area of anaesthetised animals in starlings, an extreme case of loss of attention. In Campbell's monkeys, novel and familiar species-specific and non-specific sounds (other species present or not in the environment, and non-biological sounds) were used as we expected the monkeys to respond less to sounds usually heard in their environment (e.g., horses [49]). We hypothesised that, more than a mere dominance of a hemisphere for species-specific sounds, a more subtle specialisation may be found according to the "attentional value" (e.g., "novelty") of the sound for the animal. We chose to broadcast the full series of sounds to each individual, a procedure that has proved useful in birds and may avoid controversies related to the playback of only one sound to one animal [14,50,51]. We also expected unfamiliar and familiar sounds to elicit different levels of arousal/attention that might reflect the level of reaction to the playback in terms of number of reactions and strength of orientation.

2. Results

2.1. Study 1: Behavioural Responses of Campbell's Monkeys to Familiar or Novel Sound Stimuli

This study was performed on six captive born female Campbell's monkeys, aged 4–13 years, and living in the same social group. The distractor was a homemade caramel, which was spread on the wire net inside the room just above a metal tray on which the monkeys could sit in order to lick it. This food element proved very attractive as all animals remained sitting on the tray during the whole experiment, licking it actively. Nine distinct sound categories were used that were or were not familiar or close in structure to the own species calls. These nine sound categories were: white noise (non-biological stimulus), vocalisations of familiar (conspecifics) primates and birds, and of unfamiliar male and female primates and birds. All sounds were calls with a social positive valence (for the concerned species). The term familiarity was used here as "common in the environment" and not in terms of "individual familiarity". A total of 540 playbacks were performed with 90 sounds per female (5 exemplars of 8 biological sounds × 2 + 10 times white noise). The analysis of the video recordings revealed head movements and orientations in response to playbacks that occurred within the second following the playback. Therefore, only changes in behaviour occurring in the second following playback were considered as responses.

2.1.1. Results of Study 1

Overall, a high proportion of the playbacks elicited a clear response (314 out of 540 tests, binomial test response/no response: $p = 0.0002$). No habituation could be detected as the proportion of playback eliciting a response did not differ between both sessions (Wilcoxon test, $p > 0.05$ in all cases) or according to the rendition order for white noise (Spearman rank order correlation, N = 10, r = −0.298, *p* = 0.4).

Three stimuli elicited clearly a higher proportion of responses than the others: the loud calls of the male Wolf's monkeys (63%: binomial test: *p* = 0.05), the social calls of the females Wolf's monkeys (80%: *p* = 0.0006) and the barnacle goose calls (80%: *p* = 0.0006). The most familiar sounds appeared to elicit the lowest level of response (51% for the female Campbell's monkey calls and 40% for the European starling whistles) (*p* < 0.26).

When considering individual responses in terms of lateralisation, clear differences according to stimulus were observed in the second session (Figure 1), while none was found for the first session (Wilcoxon tests, N = 6, *p* < 0.11). All subjects turned their heads significantly to the left after hearing four unfamiliar stimuli: social calls of female baboons (87% of the responses) and of female Wolf's monkeys (84%) and barnacle goose vocalisations (80%) and white noise (69%) (Wilcoxon tests, N = 6, *p* < 0.04). No significant right/left differences were found for any of the familiar stimuli, or the species-specific calls (unfamiliar in terms of individual identity but familiar in terms of overall structure: 56% for male and 53% for female Campbell's monkeys' calls elicited orientations to the right).

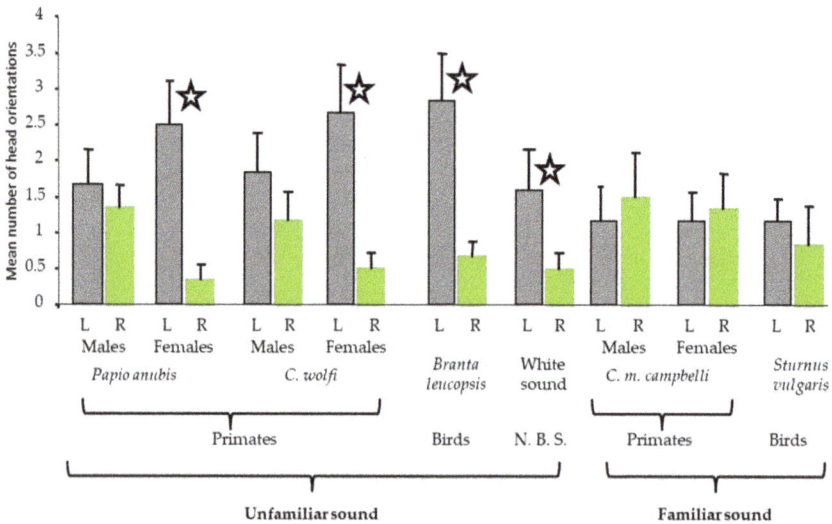

Figure 1. Variation of the mean (+/− s.e.) number of head orientations according to stimuli, calculated from all subjects. L: left orientation, R: right orientation, N.B.S.: Non-Biological Sound, Star: significant difference – Wilcoxon tests.

2.1.2. Discussion of Study 1

The broadcast of a series of sounds varying in terms of familiarity and while the animals had a food distractor showed that only novel sounds elicited responses above chance level, and most of all lateralised responses, with a right hemisphere preference. Familiar sounds, either because they were species-specific or because the species concerned were common in the environment, elicited few responses and no lateralisation. The animals' responses suggest that species-specific calls and sounds of another non-primate species common in the environment might all correspond to a same general "familiar" category, that, when animals are occupied by another preferred activity may not create enough arousal to stimulate attention. Hearing familiar sounds while focusing on an attractive "task" may not elicit arousal. Unfamiliar sounds on the other hand elicited more responses in terms of head movements, suggesting that the animals paid more attention. Another element is that these unfamiliar sounds elicited the same level of responses in the two sessions, whereas lateralisation of the responses became clear only at the second session, as if a certain level (or an increased level) of attention was necessary in order to adapt their responses. Both a probable decrease of motivation elicited by the caramel and the repetition of unfamiliar sounds could explain increased attention. In the present study, the unfamiliar sounds may have been "startling" enough to distract the monkeys from their focus on the caramel.

In another study with also a food distractor, the Campbell's monkeys confronted by familiar (group members) species-specific calls elicited lateralised responses (left head turning, right hemisphere) only when they had a negative valence (i.e., threat calls, and not for contact calls) [31]. This actually may reinforce a potential role of attention; as such signals do require an immediate arousal and readiness to respond, contrary to social positive signals. Indeed, in this same study, as in the present one, neither heterospecific calls nor social positive species-specific calls elicited any lateralised response.

The social function of the vocalisation and the subsequent attention of the female must also be taken into account. No lateralised responses to male loud calls were observed in the present study with a food distractor, whatever the familiarity level. Campbell's monkeys present strong vocal and sexual dimorphisms. Male loud calls are directed toward other males in a competitive context (or in alarm contexts, [52]) and females react much less to the loud calls of other conspecific males than to those of their harem male, which may explain why they showed no laterality for any of the male loud calls broadcasted. On the contrary, female contact calls are involved in pacifying interactions and, through vocal convergence, reflect social affinities [47,53]. In horses, lower responses and no lateralised head turning was observed when they heard the whinnies of group members in their familiar setting [54].

2.2. Study 2: Electrophysiological Responses of Auditory Neurons to Different Sounds in European Starlings

Twenty-six wild-caught adult female European starlings were used for this study. Ten were recorded while awake-restrained, and 16 while anesthetised both during and outside the breeding season. Auditory stimuli consisted in artificial non-specific sounds and songs chosen for their behavioural relevance: Class-I: species-specific whistles that are common to all males and are the bases for male-male interactions and dialectal variations; Class-II whistles that are more individual-specific but can be shared by close social (same sex) partners; and Class-III warbling motifs that are individual specific but can be shared by close social partners excepted for clicks, common in all male songs all year round and high-pitched trills that occur at the end of the warbling sequence and are more frequent at breeding time and especially in unmated males (see Section 4.2.).

2.2.1. Results of Study 2

Since there were no significant differences in terms of laterality between the recordings performed during or outside the breeding period within each category of bird (awake/anaesthetised), data were pooled. There were more neuronal responses to the auditory stimuli overall when the birds were awake than when they were anaesthetised [38].

A clear laterality of sound processing appeared in the awake-restrained birds, revealing that the individual song elements from a known bird (group member) elicited more responses in the right hemisphere (Wilcoxon test, $N = 10$, $Z = 2.54$, $p = 0.01$) while the more universal motif types (clicks and high pitched trills) elicited more responses in the left hemisphere (Wilcoxon test, $Z = 1.9$, $p = 0.05$). The artificial sounds and male-specific universal species-specific Class I whistles did not elicit any lateralised response (Figure 2).

The pattern was clearly different in the anesthetised birds, which showed no lateralisation for individual songs but a left hemisphere dominance for the artificial sounds and again for the more universal features of Class III songs (Wilcoxon tests, artificial sounds, $Z = 1.96$ $p = 0.049$; clicks and trills $Z = 2.52$, $p = 0.01$).

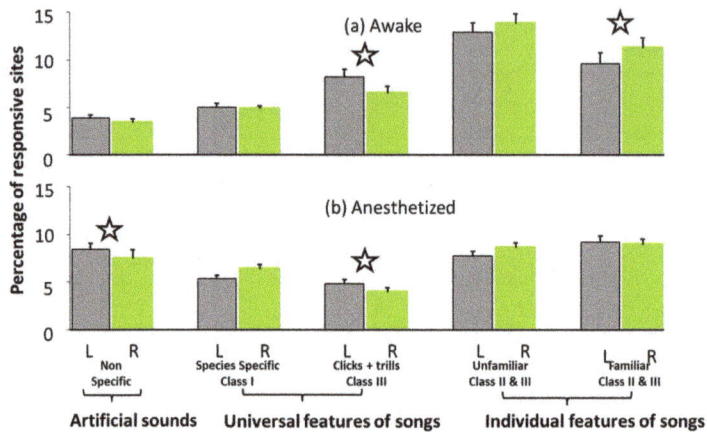

Figure 2. Laterality of neuronal preferences (%: percentage of responsive sites) in field L of awake (**a**) and anaesthetized (**b**) European Starling: percentage of neural sites that responded to nonspecific (artificial sounds), universal features of songs (species specific Class I, Class II clicks and trills) and familiar and unfamiliar individual songs (Class II whistles and Class III motifs).

2.2.2. Discussion of Study 2

The neuronal responses in field L of adult female starlings clearly differed according to the birds' wakefulness state, with more responses in the right hemisphere for a familiar bird (group member) when they were awake, a lateralised response that disappeared when they were anaesthetized. They also showed more responses towards artificial sounds when anaesthetised, these elements, similar to the more universal warbling motifs, being processed most in the left hemisphere. Interestingly, fRMI studies on anaesthetised females showed the same preference of Right Hemisphere (RH) for processing individual songs from unknown males and Left Hemisphere (LH) for the universal species-specific song elements [40].

These results differ to some degree from those obtained in male starlings [22]: in awake birds, there was an overall predominance of RH responses that we did not particularly observe in females, and the RH was more involved in processing individual songs, whether familiar or unknown. In anaesthetised birds, RH was involved in processing universal species-specific sounds while they were processed on the left in females. Such sex differences may be due to either a differential general laterality of sound processing in males and females or the fact that only male songs were broadcasted in both cases, which means they did not have the same social significance for both types of birds. Further studies involving female song should be performed.

In both sexes though, when animals are awake, individual identity is processed more in the right hemisphere while, when anaesthetised, the left hemisphere seems to be more involved in processing potentially socially meaningless sounds such as artificial sounds as is the case in anaesthetised adult female zebra finches, which also have artificial sounds processed in LH [55].

These results reinforce the idea that the right hemisphere would be more involved in processing individual information as suggested by different studies on chickens [56,57], quails [58] or sheep [59].

3. General Discussion

The results obtained on both species studied here with different paradigms (head orientation versus electrophysiological recordings) converge in showing that altered attention, either by having a distractor or through anaesthesia, leads to particular lateralised patterns of response. In the two studies using food distractor, female Campbell's monkeys, confronted by negative and positive social calls produced by conspecifics or familiar other primates, show a right hemisphere dominance for only the negatively connoted species-specific calls [31], while they also show a right hemisphere dominance for all sorts (heterospecific, artificial sounds) of novel sounds and lower non-lateralised responses for familiar sounds (including positively connoted species-specific calls).

When female starlings are awake, there is a right dominance of the Field L neurons for individual familiar songs, while, when they are anaesthetised, this dominance disappears, with LH processing potentially more meaningless sounds.

Interestingly, differences in laterality of responses in distracted animals (juice drink during playback) were also observed in two studies using two different "types of familiarity" procedures on mouse lemurs: while the authors found a LH bias for conspecific calls with a negative valence recorded in the field [33], they found no such bias according to call valence in another study where the calls were recorded from non-group members living in the same facility [30].

In the latter case, they did not find any bias for familiarity, contrary to our primate study, but they used calls from animals that were not group members but still present in the facility. Horses react to the whinnies of neighbouring (non-group members) horses as familiar compared to those produced by totally unknown horses [49,54]. Therefore, the question remains open as how the lemurs would have processed familiarity if really unknown calls had been broadcasted. Future studies on diverse species should probably differentiate between individual familiarity and sound familiarity, in terms of acoustic environment or structural proximity.

However, while in our studies both species revealed a clear influence of the attentional state on the pattern of laterality observed, they also showed apparent discrepancies. Thus, there was a RH bias for processing negatively connoted species-specific calls in distracted Campbell's monkeys, whereas RH was more involved in processing familiar rather neutral or positive familiar songs in awake starlings. In addition, while only novel (i.e., unfamiliar) sounds, whether heterospecific or artificial, elicited higher and lateralised RH responses in distracted Campbell's monkeys, starlings' neurons reacted to these sounds with a LH dominance.

This suggests that brain processing of auditory stimuli and the associated emotional valence differs between these species, that distraction and anaesthesia certainly do not represent the same level of attention alteration, or that there is another common process that may explain these discrepancies. Overall, all EEG studies converge to indicate a LH bias for positive and RH bias for negative emotional states in human studies including when processing speech, as also shown in most animal studies using visual stimuli (i.e., [58,60–64]). Animal behavioural studies on auditory perception are not as clear-cut: dogs turn more the head towards the left (RH) when hearing a thunderstorm noise or human voices with a negative valence [65,66], as do Campbell's for conspecific calls with a negative valence [31] but mouse lemurs turn the head towards the right (LH) for the same type of stimulus [33]. In addition, domestic goats show more right head turning (LH) for conspecific calls produced in supposedly negative contexts (isolation, frustration, dog barks) but also for calls produced in anticipation of feeding, a context supposedly associated with positive emotions [67]. The authors concluded that the RH processes high arousal independently of valence, although one alternative possibility is that anticipation of positive event may correspond to a quite ambiguous situation [68]. Following Baciadonna et al. [67], one hypothesis therefore would be that lateralised processes concern intensity and not valence of the stimuli, two aspects of emotions separated in the circumplex model of Lang et al. [69]. This would be in contrast to the valence theory [70] that predicts a clear impact of valence on the lateralisation of stimulus processing (see also [71,72]).

How can we explain the number of studies, including the present one, showing that individual identity, familiarity and overall functionally significant stimuli are processed with hemispheric specialisation, without any particular arousal? In dogs, fMRI studies indicate that meaningful (human) auditory stimuli are processed on the right side, while "marked" words are processed in LH [73], and behavioural studies that a "positive" human voice is processed in RH in this species [74].

Actually, there is only one way of explaining these different facets of laterality which is, as suggested by Andrew [75], attention. If, as suggested by different authors, the right hemisphere is more involved in detailed analysis (which requires attention) (e.g., [8]), then it could explain why in awake undistracted animals, it is devoted to the analysis of individual identity (e.g., [76]), in both distracted and undistracted animals to negatively connoted stimuli, that tend to attract more attention [77–80] and anticipatory contexts where the animals' attention is focused on expectation. For example, dogs processed "happy" human voices with the LH, but they also showed a lower arousal for these voices than those reflecting fear [66]. There is more activation of RH when humans watch incongruent audio-visual interactions on videos [81], and incongruence is known to stimulate attention [82]. Alertness overall tends to increase for both salient or more negative interpersonal conditions [83], which leads one to consider that the arousal elicited by auditory stimuli is more important than their specific valence. However, arousal involves attention.

When distracted by an appealing food, female Campbell's monkeys just reacted to novel and thus "incongruent" (no baboon or barnacle geese in their captive environment) sounds with left head turning (RH): familiar sounds such as contact calls of conspecifics or birds (starlings) common in their environment obviously were not salient enough to trigger reactions, and still less so laterality. Auditory neurons of male and female anaesthetised starlings showed responses to meaningless sounds, such as artificial sounds, mostly in the left hemisphere, as also observed in female zebra finches [55].

Awake female starlings showed a RH bias for particularly meaningful conspecific calls, such as the individual songs of known birds (which reflect social bonding, [43]) and distracted female Campbell's monkeys also showed a RH bias for negatively connoted calls of conspecifics [31]). One can think that the salience of the stimuli depends on the functional significance of the signals for each species. According to Syka et al. [84] and Huez et al. [85], in mammals, anaesthesia affects sensory elements that show relevance, and attention is required for processing meaningful vocalisations.

If, as proposed by Andrew and Watkins [35], we consider attention as a core aspect of hemispheric specialisation, then it would explain enhanced laterality for novelty, incongruence, and highly (e.g., socially) significant signals, as well as discrepancies between studies using (e.g., [30,65]) or not (e.g., [74]) a food distractor. Distraction may raise the threshold of attention-getting value of the stimuli and thus alter laterality.

Electrophysiological data converge to suggest a higher implication of RH in attentional processes in humans [82,86] and horses [39]. Meaningful sounds elicit more responses in the RH of dogs [73,87,88].

In the same line, Ghazanfar et al. [20] showed that rhesus monkeys oriented to the left for reversed calls, which may have been perceived as incongruent hence deserving further attention. Pohl [29] argued that the right hemisphere in baboons processes pure tones, musical sounds and vowels and he suggested that processing these unfamiliar structures is more difficult. Cynx et al. [21] and Watkins [28] proposed that the RH played a similar role in processing novel features or more complex sounds. Thus, hens' clucks are processed in the left hemisphere, but the introduction of a new note (novel feature in a familiar sound) induces processing in the right hemisphere [28].

Our results suggest, for the first time, that the attention-getting property of stimuli may be a more adapted concept for explaining hemispheric auditory specialisation (including also the species-specific vocalisations) and may explain that the distinction between familiar/novel may be more important than between species-specific versus non-specific stimuli. Thus, in our study on monkeys, the responses were clearly oriented towards the left (RH) for an unfamiliar bird (barnacle goose), whereas there was no clear orientation for another bird, the European starling, common in the animal's environment. Horses show no lateralised responses when hearing the whinnies of a group member within the familiar pasture but react with a lateralised pattern when hearing a total stranger or a familiar non-group member horse that never shares the same pasture [54].

An attention-based model may reconcile the different existing dominant hypotheses of a RH- [70,88–92], arousal/intensity or individual [56] based lateralisation, in particular for auditory perception. According to a species' social organisation/structure or life conditions, the more meaningful, hence attention-getting stimuli, may differ.

4. Materials and Methods

4.1. Study 1: Behavioural Responses of Campbell's Monkeys to Familiar or Novel Sound Stimuli

4.1.1. Subjects

This study was performed on six captive born female Campbell's monkeys, aged 4–13 years, and living in the same social group, composed of one adult male, six adult females and three juveniles (one male and two females) at the time of the experiment. The animals were housed in an enclosure divided into outdoor (21 m^2 × 4 m) and indoor (21 m^2 × 3 m) parts. Trap doors enabled the animals to move freely from one to the other part. However, during the playback sessions, they were kept indoors.

4.1.2. Procedure

The indoor part was connected to the experimental room through a trap door and a concrete wall separated both rooms. Therefore, when in the experimental room, the animals could hear but not see the rest of the group. The monkeys were trained to leave the group in order to go individually to the experimental room using food reinforcement several weeks before the onset of the experiments. The six females would go easily and did not show any sign of stress during the experiments.

The distractor was a homemade caramel, which was spread on the wire net inside the room just above a metal tray (20 cm × 20 cm) on which the monkeys could sit in order to lick it. This food element proved very attractive as all animals remained sitting on the tray during the whole experiment, licking it actively. This also ensured that during the experiments the animals would keep quietly sitting with their back towards the loudspeaker in a symmetrical position. Playbacks only occurred when the animal was in this position, licking the caramel. A video camera was placed in front of the animal in order to record its behaviour.

The experiment took place between 26 March and 25 April 2005. Two experiments were performed per day: one before food distribution in the morning and the other in the afternoon. The animals were tested individually in a rotating order and, to avoid habituation and/or loss of motivation in the situation, only two sounds were broadcast per day for a given female. Playback was manually ordered through a computer (Amiga- Commodore – U.S.A.) by the experimenter, who waited for the animal to be sitting with its ears symmetrical to the loudspeaker before starting the playback. The interval between two successive sounds was therefore variable (1–10 min) depending on the behaviour of the subject. After the first sound, the subject had to move away from the caramel and then to place itself again in front of the caramel.

4.1.3. Auditory Stimuli

Nine distinct sound categories were used that were or were not familiar or close in structure to the own species calls. These nine sound categories were: white noise (non-biological stimulus), vocalisations of familiar male and female primates, Campbell's monkeys (*C. c. campbelli*), familiar birds, European starlings (*Sturnus vulgaris*), unfamiliar male and female primates, baboons (*Papio anubis*) and Wolf's monkeys (*Cercopithecus wolfi*) and unfamiliar birds, barnacle geese (*Branta leucopsis*). Primate male calls were loud calls and female calls were contact calls. All sounds were calls with a social positive valence (for the concerned species). For each of the eight biological stimuli, calls from five distinct individuals were used to prevent pseudo-replication. The term familiarity was used here as "common in the environment" and not in terms of "individual familiarity". Thus, the species-specific calls had been recorded from wild animals unknown to the experimental animals but they were considered as familiar in terms of "category of sounds commonly heard in the environment".

The sounds were broadcast at 75 dB, as measured at 2 m (distance between the loudspeaker and the sitting tray). Each individual exemplar of sound was broadcast twice during the entire experiment: the whole series of sounds was broadcast (Session 1) before a second series of playback of the same sounds took place (Session 2). Each female therefore heard 90 sounds (5 exemplars of 8 biological sounds × 2 + 10 times white noise). A total of 540 playbacks were thus performed. The order of playbacks of the stimuli was randomised for each session and a given female never heard the same succession of two stimuli, twice.

The analysis of the video recordings revealed head movements and orientations in response to playbacks that occurred within the second following the playback, that is before any group member in the other room could produce any vocal response [25]. Therefore, only changes in behaviour occurring in the second following playback were considered as responses. Changes in head orientation were only taken into account if the head movement was above 45°. Head orientation could be left, right or none.

4.1.4. Statistical Analysis

Non-parametric statistics were used: binomial tests to compare the number of responses/non-responses, the number of left/right responses for each and all stimuli, respectively; Wilcoxon test to compare right/left responses between sessions and stimuli and to ensure inter individual validity; and Spearman rank order correlation tests to test possible correlations between playback rendition order and response rate or orientation.

4.2. Study 2: Electrophysiological Responses of Auditory Neurons to Different Sounds in European Starlings (see also [38,93])

4.2.1. Subjects

In total, 26 wild-caught adult female European starlings were used for this study. These birds had been caught as adults in October 2006 (N = 10) or 2012 (N = 16) during their autumn migration along the Normandy coast (north of France), about 3 years before the beginning of the experiments They were then kept together with other males and females caught at the same time in an outdoor aviary with food and water ad libitum. Although seasonal changes occurred in these birds (e.g., beak colour change, see below), the absence of nest boxes prevented them from breeding. Thus, females were in an appropriate seasonal environment and they showed visible seasonal characteristics, such as beak colour changes, but they were not influenced by seasonal changes in male behaviour (song, sexual display, etc.). They were brought to the laboratory and temporarily housed in single cages, with close contact (visual and auditory) with the other neighbouring birds at the beginning of the experiments.

Ten females were recorded in an awake-restrained state: six outside the breeding season, in fall (November and December 2006) and four during the breeding season in spring (April 2006).

Sixteen other females (eight in fall 2014 and eight in spring 2015) were recorded while anaesthetised using a 4 mL/kg mixture of 5 mL Medetomidine (1 mg/mL), 0.25 mL Ketamine (50 mg/mL) and 5 mL saline solution. The recordings lasted about 6 h (±10 min) and, to maintain the anaesthesia level, we injected every 2 h a third of the first dose.

The physiological state of the birds was assessed by their bill colour; yellow during the breeding season, dark during the rest of the year. This characteristic is a very good indicator of gonadal activity [94–97]. All females tested in spring had a yellow beak, indicating that they were in breeding condition, and all females tested in autumn had a black beak, indicating that they were in a non-breeding state. Prior to the neurophysiological experiments, a stainless-steel well was implanted stereotaxically on the bird's skull under halothane anaesthesia (0.4 L/min of carbogene—95% O_2/5% CO_2—saturated in halothane (2bromo2chloro1, 1, 1 trifluoroethane) and 0.6 L/min of carbogene). The centre of the implant was located precisely with reference to the bifurcation of the sagittal sinus in the left hemisphere. This position allowed the implantation of the electrodes in both hemispheres. After surgery, the birds were allowed to rest for 3 days in individual cages. During this period, they could hear but not see each other. They were kept under natural photoperiod throughout the study. During the electrophysiological recordings, the well was used for fixation of the head and as the reference electrode. Before the first recording session, the bone was removed to allow electrode introduction in both hemispheres. The bone was slightly soaked with a drop of lidocaine (4%) before removing it to avoid possible pain from the bone or the dura mater. This quantity of lidocaine was also sufficient to avoid pain during electrode insertion even if the electrode tip was too thin (diameter < 20 lm) to stimulate the dura mater nociceptors. A piece of plastic foam was placed on the bird's head between each recording session in order to protect the brain surface.

The experiments were performed in France (licence No. 005283, issued by the Department of Veterinary Services of Ille-et-Vilaine) in accordance with the European Communities Council Directive of 24 November 1986 (86/609/EEC).

4.2.2. Electrophysiological Recordings

Before the neurophysiological experiments, a stainless-steel well was implanted stereotaxically on the bird's skull under isoflurane anaesthesia (0.4 L/min of carbogene—95% O_2–5% CO_2—saturated in isoflurane and 0.6 L/min of carbogene). The centre of the implant was located precisely with reference to the bifurcation of the sagittal sinus at 2.5 mm rostral and 1 mm in the left hemisphere. This position allowed the introduction of the electrodes in both hemispheres. After implantation, the birds were allowed to rest for three days in individual cages. During this period, they could hear but not see each other. They were kept under natural photoperiod throughout the study. During the electrophysiological recordings, the well was used for head fixation and as the electrode reference. Before the first recording session, the bone was removed to allow electrode introduction in both hemispheres.

All recordings were made using the same recording setup at a temperature of about 20 °C and relative humidity of about 30%. Neuronal activity was recorded systematically throughout Field L during the broadcast of every acoustic stimulus, using the same approach as [97].

A head holder was used to maintain the bird's head in a constant and stable position. We used an array of four microelectrodes (two in each hemisphere) made of tungsten wires insulated by epoxylite (FHC). Electrode impedance was in the range of 5–6 MΩ each. These electrodes spaced 1.2 mm apart in the sagittal plane and 2 mm apart in the coronal plane. Recordings were performed in one sagittal plane in each hemisphere. These planes were precisely located with reference to the bifurcation of the sagittal sinus: 2.5 mm rostral and 1 mm in each hemisphere. These coordinates ensured that recordings were made in Field L centred on the L2 sub-area described by Capsius et al. [98] and Cousillas et al. [99]. The artificial non-specific stimuli composed by pure tones and white noise allowed us to assess the presence of the tonotopic organisation that is characteristic of Field L and to confirm that recordings were done in this area [99,100]. Recordings in the left and right hemispheres were made simultaneously, at symmetrical locations. The recording planes were at the same location for all birds. Recordings were performed at 30–40 sites along the path of an electrode penetration. Three penetrations could be done during a 6 h session. Penetrations within one recording plane were 200 μm apart. For each penetration, recordings started 600 μm below the brain surface, at a site that gave no auditory response, and continued, every 200 μm, until no response was obtained in both outermost penetrations. The dimensions of the recording plane were 2.4 mm caudo-rostral and 3.6 mm dorso-ventral (8.64 mm 2 area).

4.2.3. Auditory Stimuli

Auditory stimuli consisted in artificial non-specific sounds and songs chosen for their behavioural relevance [101]: Class-I: species-specific whistles that are common to all males and are the bases for male-male interactions and dialectal variations; Class-II whistles that are more individual-specific but can be shared by close social (same sex) partners; and Class-III warbling motifs that are individual specific but can be shared by close social partners excepted for clicks, common in all male songs all year round and high-pitched trills that occur at the end of the warbling sequence and are more frequent at breeding time and especially in unmated males [102,103].

The stimulus set was made of these artificial non-specific sounds and exemplars of the three classes of songs (Figure 3). Although no adaptation was reported in the Field L using this kind of stimulus set [100], the stimuli were broadcast with intervals of at least 300 ms in order to avoid any problem of adaptation between the stimuli. The sequence of stimuli set was determined randomly and then the same sequence was repeated 10 times at each recording site.

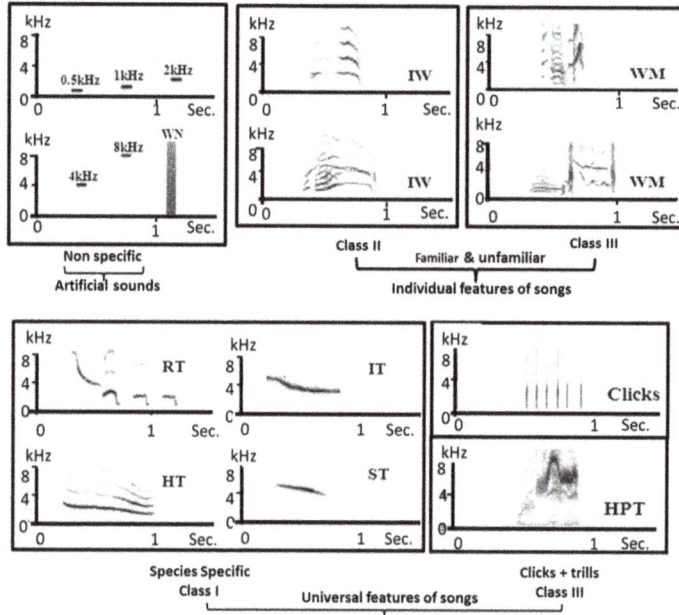

Figure 3. Stimulus set: artificial non-specific sounds and exemplars of Class I (species specific whistles), Class II (individual whistles) and Class III (individual warbling motifs and species specific clicks and trills)

Spike arrival times were obtained (with a temporal resolution of 0.1 ms) by thresholding the extra-cellular recordings with a custom-made time- and level-window discriminator [97]. Single units or small multiunit clusters of 2–4 neurons were recorded in this manner. Since several studies found that analyses resulting from single and multi-units led to similar results [104,105], the data from both types of units were analysed together.

The computer that delivered the stimuli also recorded the times of action potentials and displayed on-line rasters of the spike data for the four electrodes simultaneously. At each recording site, spontaneous activity was measured for 1.55 s before the presentation of the first stimulus of each sequence, which resulted in 10 samples of spontaneous activity (i.e., a total of 15.5 s).

Neuronal responsiveness was assessed as in George et al. [23] by comparing activity level (number of action potentials) during stimulation and spontaneous activity using binomial tests. Only responsive sites were further analysed by calculating the proportion of sites responding to each stimulus and to each class of stimuli. The mean values calculated for individual birds were then used for statistical comparisons.

4.2.4. Statistical Analyses

Given the low number of subjects, non-parametric statistics were used to test for potential differences between the two hemispheres.

Author Contributions: Conceptualisation, M.H., H.C., I.G. and C.B.-H.; Data curation, A.M., G.K. and I.G.; Formal analysis, M.H., A.M. and C.B.-H.; Funding acquisition, M.H.; Investigation, A.M. and C.B.-H.; Methodology, M.H., H.C., A.M. and C.B.-H.; Supervision, M.H. and C.B.-H.; and Writing—original draft, M.H., H.C., A.L. and C.B.-H.

Funding: This research received no external funding

Acknowledgments: We thank A. Rossard and P. Bec for taking care of the Paimpont primates, and D-M Paquet and S. Alcaix for electrophysiological recordings.

Conflicts of Interest: The authors declare no conflict of interest.

References

1. Broca, P. Remarques sur le siège de la faculté du langage articulé suivies d'une observation d'aphémie. *Bull. Soc. Anat.* **1861**, *6*, 398–407.

2. Rogers, L.J.; Andrew, R.J. *Comparative Vertebrate Lateralization*; Cambridge University Press: Cambridge, UK, 2002; ISBN 0-521-78161-2.

3. Mc Neilage, P.F.; Rogers, L.J.; Vallortigara, G. Origins of left and right Brain. *Sci. Am.* **2009**, *301*, 60–67. [CrossRef]

4. Vallortigara, G.; Chiandetti, C.; Sovrano, V.A. Brain asymmetry (animal). *Adv. Rev.* **2011**, *2*, 146–157. [CrossRef] [PubMed]

5. Rogers, L.; Vallortigara, G. When and why did brains break symmetry? *Symmetry* **2015**, *7*, 2181–2194. [CrossRef]

6. Taglialatela, J.P. Functional and structural asymmetries for auditory perception and vocal production in nonhuman primates. *Spec. Top. Primatol.* **2007**, *5*, 120–145.

7. Ocklenburg, S.; Ströckens, F.; Güntürkün, O. Lateralisation of conspecific vocalization in non-human vertebrates. *Laterality* **2013**, *18*, 1–31. [CrossRef]

8. George, I. Hemispheric asymmetry of songbirds. In *The Two Halves of the Brain: Information Processing in the Cerebral Hemispheres*; MIT Press Scholarship Online; Hugdahl, K., Westerhausen, R., Eds.; MIT Press: Cambridge, MA, USA, 2010; pp. 91–120. [CrossRef]

9. Konerding, W.S.; Zimmermann, E.; Bleich, E.; Hedrich, H.J.; Scheumann, M. The head turn paradigm to assess auditory laterality in cats: Influence of ear position and repeated sound presentation. *PeerJ* **2017**, *5*, e3925. [CrossRef]

10. Böye, M.; Güntürkün, O.; Vauclair, J. Right ear advantage for conspecific calls in adults and subadults, but not infants, California sea lions (*Zalophus californianus*): Hemispheric specialization for communication? *Eur. J. Neurosci.* **2005**, *21*, 1727–1732. [CrossRef]

11. Ehret, G. Left hemisphere advantage in the mouse brain for recognizing ultrasonic communications calls. *Nature* **1987**, *325*, 249–251. [CrossRef]

12. Palleroni, A.; Hauser, M. Experience-dependant plasticity for auditory processing in a raptor. *Science* **2003**, *299*, 1195. [CrossRef]

13. Siniscalchi, M.; Laddago, S.; Quaranta, A. Auditory lateralization of conspecific and heterospecific vocalizations in cats. *Laterality* **2016**, *21*, 215–227. [CrossRef] [PubMed]

14. Hauser, M.D.; Andersson, K. Left hemisphere dominance for processing vocalizations in adult, but no infant, rhesus monkeys: Field experiments. *Proc. Natl. Acad. Sci. USA* **1994**, *91*, 3946–3948. [CrossRef] [PubMed]

15. Hauser, M.D.; Agnetta, B.; Perez, C. Orienting asymmetries in rhesus monkeys: The effect on time-domain changes on acoustic perception. *Anim. Behav.* **1998**, *56*, 41–47. [CrossRef] [PubMed]

16. Szymanska, J.; Trojan, M.; Jakucinska, A.; Wejchert, K.; Kapusta, M.; Sikorska, J. Brain functional asymmetry of chimpanzees (*Pan troglodytes*): The example of auditory laterality. *Pol. Psychol. Bull.* **2017**, *48*, 87–92. [CrossRef]

17. Heffner, H.E.; Heffner, R.S. Temporal lobe lesions and perception of species-specific vocalizations by macaques. *Science* **1984**, *236*, 75–76. [CrossRef]

18. Heffner, H.E.; Heffner, R.S. Effect of unilateral and bilateral auditory cortex lesions on the discrimination of vocalizations by Japanese macaques. *J. Neurophysiol.* **1986**, *56*, 683–701. [CrossRef] [PubMed]

19. Okanoya, K.; Ikebuchi, M.; Uno, H.; Watanabe, S. Left-side dominance for song discrimination in Bengalese finches (*Lonchura striata var. domestica*). *Anim. Cogn.* **2001**, *4*, 241–245. [CrossRef]

20. Ghazanfar, A.A.; Smith-Rohrberg, D.; Hauser, M.D. The Role of Temporal Cues in Rhesus Monkey Vocal Recognition: Orienting Asymmetries to Reversed Calls. *Brain Behav. Evol.* **2001**, *58*, 163–172. [CrossRef]

21. Cynx, J.; Williams, H.; Nottebohm, F. Hemispheric differences in avian song discrimination. *Proc. Natl. Acad. Sci. USA* **1992**, *89*, 1372–1375. [CrossRef]

22. George, I.; Vernier, B.; Richard, J.-P.; Hausberger, M.; Cousillas, H. Hemispheric specialization in the primary auditory area of awake and anesthetized starlings. *Behav. Neurosci.* **2004**, *118*, 597–610. [CrossRef]

23. George, I.; Cousillas, H.; Richard, J.-P.; Hausberger, M. State-dependent hemispheric specialization in the songbird brain. *J. Comp. Neurol.* **2005**, *18*, 48–60. [CrossRef] [PubMed]

24. Gil-da-Costa, R.; Hauser, M.D. Vervet monkeys and humans show brain asymmetries for processing conspecific vocalizations, but with opposite patterns of laterality. *Proc. R. Soc. B* **2006**, *273*, 2313–2318. [CrossRef] [PubMed]

25. Lemasson, A.; Koda, H.; Kato, A.; Oyakawa, C.; Blois-Heulin, C.; Masataka, N. Influence of sound specificity and familiarity on Japanese macaques' (*Macaca fuscata*) auditory laterality. *Behav. Brain Res.* **2010**, *208*, 286–289. [CrossRef] [PubMed]

26. Arve, E.; Hugdahl, K. Attentional effect in dichotic listening. *Brain and language* **1995**, *49*, 189–201.

27. Locke, J.L.; Snow, C. Social influences on vocal learning in human and non-human primates. In *Social Influences on Vocal Development*; Snowdon, C.T., Hausberger, M., Eds.; Cambridge University Press: Cambridge, UK, 1997; pp. 274–292.

28. Watkins, J.A.S. Lateralization of Auditory Learning and Processing in the Domestic Chick (*Gallus gallus domesticus*). Ph.D. Thesis, University of Sussex, Sussex, UK, 1999.

29. Pohl, P. Central auditory processing. V: Ear advantages acoustic stimuli in Baboons. *Brain Lang.* **1983**, *20*, 44–53. [CrossRef]

30. Leliveld, L.M.C.; Scheumann, M.; Zimmermann, E. Effects of caller characteristics on auditory laterality in an early primate (*Microcebus murinus*). *PLoS ONE* **2010**, *5*, e9031. [CrossRef]

31. Basile, M.; Lemasson, A.; Blois-Heulin, C. Social and Emotional Values of Sounds Influence Human (*Homo sapiens*) and Non-human Primate (*Cercopithecus campbelli*) Auditory Laterality. *PLoS ONE* **2009**, *4*, e6295. [CrossRef]

32. Xue, F.; Fang, G.; Yang, P.; Zhao, E.; Brauth, S.E.; Tang, Y. The biological significance of acoustic stimuli determines ear preference in the music frog. *J. Exp. Biol.* **2015**, *218*, 740–747. [CrossRef]

33. Scheumann, M.; Zimmermann, E. Sex-specific asymmetries I communication sound perception are not related to hand preference in an early primate. *BMC Biol.* **2008**, *6*, 3. [CrossRef]

34. Reinholz-Trojan, A.; Włodarczyk, E.; Trojan, M.; Kulcz Nski, A.; Stefá Nska, J. Hemispheric specialization in domestic dogs *Canis familiaris* for processing different types of acoustic stimuli. *Behav. Proc.* **2012**, *91*, 202–205. [CrossRef]

35. Andrew, R.J.; Watkins, J.A.S. Evidence for cerebral lateralization from senses other than vision. In *Comparative Vertebrate Lateralization*; Rogers, L.J., Andrew, R.J., Eds.; Cambridge University Press: Cambridge, UK, 2002; pp. 365–382. ISBN 0-521-78161-2.

36. Chanvallon, S.; Blois-Heulin, C.; de Latour, P.R.; Lemasson, A. Spontaneous approaches of divers by free-ranging orcas (*Orcinus orca*): Age- and sex-differences in exploratory behaviours and visual laterality. *Sci. Rep.* **2017**, *7*, 10922. [CrossRef] [PubMed]

37. Vallortigara, G.; Rogers, L.J.; Bisazza, A. Possible evolutionary origins of cognitive brain lateralization. *Brain Res. Rev.* **1999**, *30*, 164–175. [CrossRef]

38. Karino, G.; George, I.; Loison, L.; Heyraud, C.; De Groof, G.; Hausberger, M.; Cousillas, H. Anesthesia and brain sensory processing: Impact on neuronal responses in a female songbird. *Sci. Rep.* **2016**, *6*, 39143. [CrossRef] [PubMed]

39. Rochais, C.; Sébilleau, M.; Menoret, M.; Oger, M.; Henry, S.; Hausberger, M.; Cousillas, H. Attentional state and brain processes: State-dependent lateralization of EEG profiles in horses. *Sci. Rep.* **2018**, *8*, 10153. [CrossRef] [PubMed]

40. De Groof, G.; Poirier, C.; George, I.; Hausberger, M.; Van der Linden, A. Functional changes between seasons in the male songbird auditory forebrain. *Front. Behav. Neurosci.* **2013**, *7*, 196. [CrossRef] [PubMed]

41. Lemasson, A.; Zuberbühler, K.; Hausberger, M. Socially meaningful vocal plasticity in Campbell's monkeys. *J. Comp. Psychol.* **2005**, *119*, 220–229. [CrossRef]

42. Lemasson, A.; Glas, L.; Barbu, S.; Lacroix, A.; Guilloux, M.; Remeuf, K.; Koda, H. Youngsters do not pay attention to conversational rules: Is this so for nonhuman primates? *Sci. Rep.* **2011**, *1*, 22. [CrossRef]

43. Hausberger, M.; Foraste, M.; Richard-Yris, M.-A.; Nygren, C. Differential response of female starlings to shared and nonshared song types. *Etologia* **1997**, *5*, 31–38.

44. George, I.; Richard, J.-P.; Cousillas, H.; Hausberger, M. No need to Talk, I Know You: Familiarity Influences Early Multisensory Integration in a Songbird's Brain. *Front. Behav. Neurosci.* **2011**, *4*, 193. [CrossRef]

45. Hubel, D.H.; Henson, C.O.; Rupert, A.; Galambos, R. "Attention" Units in the Auditory Cortex. *Science* **1959**, *129*, 1279–1280. [CrossRef]

46. Henry, L.; Bourguet, C.; Coulon, M.; Aubry, C.; Hausberger, M. Sharing mates and nestboxes is associated with female 'friendship' in European starlings *Sturnus vulgaris*. *J. Comp. Psychol.* **2013**, *157*, 1–13. [CrossRef] [PubMed]

47. Lemasson, A.; Hausberger, M. Patterns of vocal sharing and social dynamics in a Campbell's monkeys. *J. Comp. Psychol.* **2004**, *118*, 347–359. [CrossRef] [PubMed]

48. Lemasson, A.; Blois-Heulin, C.; Jubin, R.; Hausberger, M. Female social relationships in a captive group of Campbell's monkeys (*Cercopithecus campbelli campbelli*). *Am. J. Prim.* **2006**, *68*, 1161–1170. [CrossRef] [PubMed]

49. Lemasson, A.; Boutin, A.; Boivin, S.; Blois-Heulin, C.; Hausberger, M. Horse (*Equus caballus*) whinnies: A source of social information. *Anim. Cogn.* **2009**, *12*, 693–704. [CrossRef]

50. Rogers, L.J. Evolution of hemispheric specialization: Advantages and disadvantages. *Brain Lang.* **2000**, *73*, 236–253. [CrossRef] [PubMed]

51. Hopkins, W.D.; Fernandez Carriba, S. Laterality of communicative behaviours in Non-human Primates: A critical analysis. In *Comparative Vertebrate Lateralization*; Rogers, L.J., Andrew, R.J., Eds.; Cambridge University Press: Cambridge, UK, 2002; pp. 445–479. ISBN 0-521-78161-2.

52. Ouattara, K.; Lemasson, A.; Zuberbühler, K. The alarm calls system of female Campbell's monkeys. *Anim. Behav.* **2009**, *78*, 35–44. [CrossRef]

53. Lemasson, A.; Gandon, E.; Hausberger, M. Attention to elders' voice in non-human primates. *Biol. Lett.* **2010**. [CrossRef]

54. Basile, M.; Boivin, S.; Boutin, A.; Blois-Heulin, C.; Hausberger, M.; Lemasson, A. Socially dependent auditory laterality in domestic horses *Equus caballus*. *Anim. Cogn.* **2009**, *12*, 611–619. [CrossRef]

55. Hauber, M.E.; Cassey, P.; Woolley, S.M.; Theunissen, F.E. Neurophysiological response selectivity for conspecific songs over synthetic sounds in the auditory forebrain of non-singing female songbirds. *J. Comp. Physiol. A* **2007**, *193*, 765–774. [CrossRef]

56. Vallortigara, G.; Andrew, R.J. Laterality of response by chicks to change in a model partner. *Anim. Behav.* **1991**, *41*, 187–194. [CrossRef]

57. Deng, C.; Rogers, L.J. Social recognition and approach in the chick: Laterality and effect of visual experience. *Anim. Behav.* **2002**, *63*, 697–706. [CrossRef]

58. Zucca, P.; Sovrano, V.A. Animal lateralization and social recognition: Quails use their visual hemifield when approaching a companion and their right visual hemi-field when approaching a stranger. *Cortex* **2008**, *44*, 13–20. [CrossRef] [PubMed]

59. Peirce, J.W.; Leigh, A.E.; Kendrick, K.M. Configurational coding, familiarity and the right hemisphere advantage for face recognition in sheep. *Neuropsychologia* **2000**, *38*, 475–483. [CrossRef]

60. Siniscalchi, M.; Quaranta, A.; Rogers, L.J. Hemispheric specialization in dogs for processing different acoustic stimuli. *PLoS ONE* **2008**, *3*, e3349. [CrossRef] [PubMed]

61. Siniscalchi, M.; d'Ingeo, S.; Minunno, M.; Quaranta, A. Communication in dogs. *Animals* **2018**, *8*, 131. [CrossRef]

62. Baciadonna, L.; Nawroth, C.; Briefer, E.F.; McElligott, A.G. Perceptual lateralization of vocal stimuli in goats. *Curr. Zool.* **2018**, *3*, 1–8. [CrossRef]

63. Lang, P.J.; Greenwald, M.K.; Bradley, M.M.; Hamm, A.O. Looking at pictures: Affective, facial, visceral, and behavioral reactions. *Psychophysiology* **1993**, *30*, 261–273. [CrossRef]

64. Andics, A.; Gácsi, M.; Faragó, T.; Kis, A.; Miklósi, A. Report voice-sensitive regions in the dog and human brain are revealed by comparative fMRI. *Curr. Biol.* **2014**, *24*, 574–578. [CrossRef]

65. Ratcliffe, V.F.; Reby, D. Orienting asymmetries in dogs' responses to different communicatory components of human speech. *Curr. Biol.* **2014**, *24*, 2908–2912. [CrossRef]

66. Andrew, R.J. The differential roles of right and left sides of the brain in memory formation. *Behav. Brain Res.* **1999**, *98*, 289–295. [CrossRef]

67. Vallortigara, G.; Andrew, R.J. Differential involvement of right and left hemisphere in individual recognition in the domestic chick. *Behav. Proc.* **1994**, *33*, 41–57. [CrossRef]

68. Austin, N.P.; Rogers, L.J. Asymmetry of flight and escape turning responses in horses. *Laterality* **2007**, *12*, 464–474. [CrossRef] [PubMed]

69. Blois-Heulin, C.; Crevel, M.; Böye, M.; Lemasson, A. Visual laterality in dolphins: Importance of the familiarity of stimuli. *BMC Neurosci.* **2012**, *13*, 2–8. [CrossRef] [PubMed]

70. De Boyer Des Roches, A.; Richard-Yris, M.A.; Henry, S.; Ezzaouïab, M.; Hausberger, M. Laterality and emotions: Visual laterality in the domestic horse (*Equus caballus*) differs with objects' emotional value. *Physiol. Behav.* **2008**, *94*, 487–490. [CrossRef]

71. Larose, C.; Rogers, L.J.; Richard, M.A.; Hausberger, M. Laterality of horses associated with emotionality in novel situations. *Laterality* **2006**, *11*, 355–367. [CrossRef] [PubMed]

72. Smith, A.V.; Proops, L.; Grounds, K.; Wathan, J.; McComb, K. Functionally relevant responses to human facial expressions of emotion in the domestic horse (*Equus caballus*). *Biol. Lett.* **2016**, *12*, 20150907. [CrossRef]

73. Zimmerman, P.H.; Buijs, S.A.F.; Bolhuis, J.E.; Keeling, L.J. Behaviour of domestic fowl in anticipation of positive and negative stimuli. *Anim. Behav.* **2011**, *81*, 569–577. [CrossRef]

74. Davidson, R.J. Emotion and affective style: Hemispheric substrates. *Psychol. Sci.* **1992**, *3*, 39–43. [CrossRef]

75. Quaranta, A.; Siniscalchi, M.; Vallortigara, G. Asymetric tail-wagging responses by dogs to different emotive stimuli. *Curr. Biol.* **2007**, *117*, 199–201. [CrossRef]

76. Siniscalchi, M.; Lusito, R.; Vallortigara, G.; Quaranta, A. Seeing left- or right-asymmetric tail wagging produces different emotional responses in dogs. *Curr. Biol.* **2013**, *23*, 2279–2282. [CrossRef]

77. Armony, J.L.; Dolan, R.J. Modulation of spatial attention by fear-conditioned stimuli: An event-related fMRI study. *Neuropsychologia* **2002**, *40*, 817–826. [CrossRef]

78. Fox, E.; Russo, R.; Dutton, K. Attentional bias for threat: Evidence for delayed disengagement from emotional faces. *Cogn. Emot.* **2002**, *16*, 355–379. [CrossRef] [PubMed]

79. Smith, A.V.; Proops, L.; Grounds, K.; Wathan, J.; Scott, S.K.; McComb, K. Domestic horses (*Equus caballus*) discriminate between negative and positive human nonverbal vocalizations. *Sci. Rep.* **2018**, *8*, 13052. [CrossRef] [PubMed]

80. Welp, T.; Rushen, J.; Kramer, D.L.; Festa-Bianchet, M.; de Passillé, A.M.B. Vigilance as a measure of fear in dairy cattle. *Appl. Anim. Behav. Sci.* **2004**, *87*, 1–13. [CrossRef]

81. Balconi, M.; Vanutelli, M.E. Vocal and visual stimulation, congruence and lateralization affect brain oscillations in interspecies emotional positive and negative interactions. *J. Soc. Neuro.* **2016**, *11*, 297–310. [CrossRef]

82. Proops, L.; Mc Comb, K. Cross-modal individual recognition in domestic horses (*Equus caballus*) extends to familiar humans. *Proc. R. Soc. B* **2012**, *279*, 3131–3138. [CrossRef] [PubMed]

83. Karakas, S.; Erzengin, Ö.U.; Başar, E. A new strategy involving multiple cognitive paradigms demonstrates that ERP components are determined by the superposition of oscillatory responses. *Clin. Neurophysiol.* **2000**, *111*, 1719–1732. [CrossRef]

84. Syka, J.; Kuta, D.; Popelar, J. Responses to species-specific vocalizations in the auditory cortex of awake and anesthetized guinea pigs. *Hear. Res.* **2005**, *206*, 177–184. [CrossRef]

85. Huetz, C.; Philibert, B.; Edeline, J.M. A spike-timing code for discriminating conspecific vocalizations in the thalamocortical system of anesthetized and awake guinea pigs. *J. Neurosci.* **2009**, *29*, 334–350. [CrossRef]

86. Ishii, R.; Canuet, L.; Ishihara, T.; Aoki, Y.; Ikeda, S.; Hata, M.; Katsimichas, T.; Gunji, A.; Takahashi, H.; Nakahachi, T.; et al. Frontal midline theta rhythm and gamma power changes during focused attention on mental calculation: An MEG beam former analysis. *Front. Hum. Neurosci.* **2014**, *8*, 1–10. [CrossRef]

87. Andics, A.; Gábor, A.; Gácsi, M.; Faragó, T.; Szabó, D.; Miklósi, Á. Neural mechanisms for lexical processing in dogs. *Science* **2016**, *353*, 1030–1032. [CrossRef] [PubMed]

88. Bach, J.-P.; Lüpke, M.; Dziallas, P.; Wefstaedt, P.; Uppenkamp, S.; Seifert, H.; Nolte, I. Auditory functional magnetic resonance imaging in dogs—Normalization and group analysis and the processing of pitch in the canine auditory pathways. *BMC Vet. Res.* **2016**, *12*, 32. [CrossRef] [PubMed]

89. Borod, J.C.; Koff, E.; Caron, H.S. Right hemispheric specialization for the expression and appreciation of emotion; a focus on face. In *Cognitive Processes in the Right Hemisphere*; Perecman, E., Ed.; Academic Press: New York, NY, USA, 1983.

90. Tucker, D.M. Lateral brain function, emotion and conceptualization. *Psychol. Bull.* **1981**, *89*, 19–46. [CrossRef] [PubMed]

91. Siberman, E.K.; Weingarten, H. Hemispheric lateralization of function related to emotion. *Brain Cogn.* **1986**, *5*, 322–353. [CrossRef]

92. Sackeim, H.; Gur, R.C. Lateral asymmetry in intensity of emotional expression. *Neuropsychologia* **1978**, *16*, 473–481. [CrossRef]
93. Cousillas, H.; George, I.; Alcaix, S.; Henry, L.; Richard, J.P.; Hausberger, M. Seasonal female brain plasticity in processing social vs. sexual vocal signals. *Eur. J. Neurosci.* **2013**, *37*, 728–734. [CrossRef] [PubMed]
94. Dawson, A. Plasma gonadal steroid levels in wild starlings (*Sturnus vulgaris*) during the annual cycle and in relation to the stages of breeding. *Gen. Comp. Endocrinol.* **1983**, *49*, 286–294. [CrossRef]
95. Ball, G.F.; Wingfield, J.C. Changes in plasma levels of luteinizing hormone and sex steroid hormones in relation to multiple-broodedness and nest-site density in male starlings. *Physiol. Zool.* **1987**, *60*, 191–199. [CrossRef]
96. De Ridder, E.; Pinxten, R.; Mees, V.; Eens, M. Short- and longterm effects of male-like concentrations of testosterone on female European starlings (*Sturnus vulgaris*). *Auk* **2002**, *119*, 487–497. [CrossRef]
97. George, I.; Cousillas, H.; Richard, J.-P.; Hausberger, M. A new extensive approach to single-unit responses using multisite recording electrodes: Application to the songbird brain. *J. Neurosci. Methods* **2003**, *125*, 65–71. [CrossRef]
98. Capsius, B.; Leppelsack, H.J. Response patterns and their relationship to frequency analysis in auditory forebrain centers of a songbird. *Hear. Res.* **1999**, *136*, 91–99. [CrossRef]
99. Cousillas, H.; Leppelsack, H.J.; Leppelsack, E.; Richard, J.P.; Mathelier, M.; Hausberger, M. Functional organization of the forebrain auditory centers of the European starling. A study based on natural sounds. *Hear. Res.* **2005**, *207*, 10–21. [CrossRef] [PubMed]
100. Leppelsack, H.J.; Vogt, M. Responses of auditory neurons in the forebrain of a songbird to stimulation with species-specific sounds. *J. Comp. Physiol.* **1976**, *107*, 263–274. [CrossRef]
101. Hausberger, M. Social influences on song acquisition and sharing in the European starling (*Sturnus vulgaris*). In *Social Influences on Vocal Development*; Snowdon, C.T., Hausberger, M., Eds.; Cambridge University Press: Cambridge, UK, 1997; pp. 128–156.
102. Verheyen, R.F. Breeding strategies of the starling. In *Bird Problems in Agriculture*; Wright, E.N., Inglis, I.R., Feare, C.J., Eds.; British Crop. Protection Council: Croydon, UK, 1980; pp. 69–82.
103. Henry, L.; Hausberger, M.; Jenkins, P.F. The use of song repertoire changes with pairing status in male European starling. *Bioacoustics* **1994**, *5*, 261–266. [CrossRef]
104. Amin, N.; Grace, J.A.; Theunissen, F.E. Neural response to bird's own song and tutor song in the zebra finch field L and caudal mesopallium. *J. Comp. Physiol.* **2004**, *190*, 469–489. [CrossRef]
105. Grace, J.A.; Amin, N.; Singh, N.C.; Theunissen, F.E. Selectivity for conspecific song in the zebra finch auditory forebrain. *J. Neurophysiol.* **2003**, *89*, 472–487. [CrossRef] [PubMed]

![symmetry logo] *symmetry*

MDPI

Article

Dynamics of Laterality in Lake Tanganyika Scale-Eaters Driven by Cross-Predation

Michio Hori [1,*], Masanori Kohda [2], Satoshi Awata [2] and Satoshi Takahashi [3]

[1] Animal Ecology, Department of Zoology, Graduate School of Science, Kyoto University, Kitashirakawa-Oiwakecho, Sakyo 606-8502, Japan

[2] Animal Sociology, Department of Biology and Geosciences, Graduate School of Science, Osaka City University, 3-3-138 Sugimoto, Sumiyoshi, Osaka 558-8585, Japan; maskohda@sci.osaka-cu.ac.jp (M.K.); sa-awata@sci.osaka-cu.ac.jp (S.A.)

[3] Graduate School of Humanities and Sciences, Nara Women's University, Nara 630-8506, Japan; takahasi@lisboa.ics.nara-wu.ac.jp

* Correspondence: hori@terra.zool.kyoto-u.ac.jp; Tel.: +81-75-753-4092; Fax: +81-75-753-4100

Received: 24 November 2018; Accepted: 14 January 2019; Published: 20 January 2019

Abstract: Scale-eating cichlid fishes, *Perissodus* spp., in Lake Tanganyika have laterally asymmetrical bodies, and each population is composed of righty and lefty morphs. Righty morphs attack the right side of prey and lefty morphs do the opposite. This anti-symmetric dimorphism has a genetic basis. Temporal changes in the frequencies of morphs in two cohabiting scale-eating species (*Perissodus microlepis* and *P. straeleni*) were investigated over a 31-year period on a rocky shore at the southern end of the lake. Dimorphism was maintained dynamically during the period in both species, and the frequencies oscillated with a period of about four years in a semi-synchronized manner. Recent studies have indicated that this type of anti-symmetric dimorphism is shared widely among fishes, and is maintained by frequency-dependent selection between predator and prey species. The combinations of laterality in each scale-eater and its victim were surveyed. The results showed that "cross-predation", in which righty predators catch lefty prey and lefty predators catch righty prey, occurred more frequently than the reverse combination ("parallel-predation"). The cause of the predominance of cross-predation is discussed from the viewpoint of the physical and sensory abilities of fishes.

Keywords: scale-eater; *Perissodus*; lateral dimorphism; frequency-dependent selection; cross-predation

---

1. Introduction

Lateral bias of behavior has been observed in various animals, particularly vertebrates. Among vertebrates, fish are expected to exhibit primordial forms of vertebrate laterality, as they are the most ancestral group. Numerous studies have used fish to elucidate the causes of behavioral laterality, and many have demonstrated an association between behavioral laterality and brain lateralization [1–8]. Many studies on morphological asymmetry have been conducted in both invertebrates and vertebrates including fish, from which three types of asymmetry have been categorized based on the frequency distribution of measured laterality: fluctuating asymmetry (unimodal and symmetrical distribution), directional asymmetry (unimodal distribution shifted from symmetry) and anti-symmetry (bimodal distribution) [9]. The relationship between behavioral laterality and morphological asymmetry, especially for anti-symmetry, had rarely been investigated aside from asymmetry of the brain [6,10,11].

Some theoretical and/or empirical studies have investigated the mechanism that is responsible for maintaining the lateralization in one population from the viewpoint of cerebral lateralization [12–14], but few studies have been done to analyze the mechanism that maintains the anti-symmetric dimorphism in one population. In particular, the population in which the dimorphism is maintained dynamically, i.e., the frequency of righty and lefty morphs changes temporally with in a fixed range, has not been investigated except for scale-eating cichlid fish in Lake Tanganyika [15]. Note that, if the frequency of righty and lefty morphs is changeable, it is not laterality at the population level, even though the laterality is obvious at the individual level.

The scale-eating cichlid fishes, *Perissodus* spp., in Lake Tanganyika have laterally asymmetric bodies, and each population is composed of righty and lefty individuals or morphs [15–17]. Righty individuals attack the right side of prey and lefty fish do the opposite [15,18,19]. Similar morphological and behavioral laterality was also observed in an Amazonian characid scale-eater, *Exodon* sp. [20]. This type of dimorphism has a genetic basis [15,21–24]. The ratio of laterality (frequency of righty morphs in each population) of *P. microlepis* oscillates around 0.5 with a period of five years, and this balance appears to be maintained by frequency-dependent selection (minority advantage) mediated by the vigilance of prey [15,25,26].

A field study of the balance of polymorphism in *P. microlepis* was conducted on rocky shores at the northern end of the lake (near Uvira, Democratic Republic of the Congo), where *P. microlepis* was essentially the only species of scale-eater in the fish community [15]. From this study, further questions arose, such as how laterality interacts in a locality where two species of scale-eater coexist, and whether polymorphism is maintained in the two species.

In order to know the temporal change of the ratio of laterality where two species of scale-eater cohabit, we monitored the ratios of their laterality on a rocky shore at the southern end of the lake (near Mpulungu, Republic of Zambia) for 31 years from 1988 to 2018, where *P. microlepis* and *P. straeleni* cohabited [27,28]. The two species (Figure 1) are the sister species in phylogeny [29], and occupy the same feeding niche, i.e., they attack nearly the same species of cichlid fishes (mainly algal-feeders with high body depth such as *Petrochromis* spp., *Cyathopharynx furcifer* and *Tropheus moorii*) as prey of their scale-eating [17,30].

Figure 1. The laterality of two scale-eating cichlids, *Perissodus straeleni* (**top**) and *Perissodus microlepis* (**bottom**) in Lake Tanganyika. A lefty morph of the former and righty morph of the latter species are shown from both sides.

Another purpose of the present study was to investigate the relation between the laterality of scale-eater and that of their prey fish. Recently Hori et al. [31] demonstrated that almost all fishes have the same kind of laterality at various intensities in a similar way to that of scale-eaters, which had been suggested for various types of fish [21,32–36]. Given the findings that all fishes have laterality, we have to reconsider the mechanism responsible for maintaining that laterality and for driving the oscillation of the ratio of laterality. The most probable mechanism may still be the prey—predator interactions, but it has come to light that the mechanism is embedded in the genetic ability of both predator and prey as well as the prey's switching of vigilance toward the majority morph of predator. Therefore, we have to examine the combination of laterality between each predator and its prey.

In the situation that all the fishes have laterality, two types of predation incident can be distinguished: A predator catches a prey of the same morph of laterality (righty predator catches righty prey and lefty one lefty prey) or a predator catches a prey of the opposite morph (righty predator catches lefty prey and lefty one righty prey). The former type can be called "parallel-predation" and the latter "cross-predation" [32,36–38]. Then, provided that the ratios of laterality of both predator and prey are maintained in any pattern, we can predict that the predation incidents in a community as a whole at any one time are biased toward an excess of one type of predation over the other, i.e., either that cross-predations are prominent over parallel-predation (predominance of cross-predation) or the reverse (predominance of parallel-predation), but not that both types of predation occur at a similar frequency (random-predation).

To verify this prediction, during the latter half of the field work (from 2006 to 2018) we observed the scale-eating in water and tried to collect both the scale-eater and its victim fish just after the scale-eating took placed, and examined the frequencies of predation types among the sampled pairs.

In short, the purpose of this study is to demonstrate the long-term dynamics of laterality in the two co-habiting species of scale-eater in one locality, and also detect the predominance of either cross predation or parallel-predation in the events of scale-eating. Some discussion is made on the meaning of the pattern found.

2. Materials and Methods

2.1. Sampling of Scale-Eaters to Assess the Ratios of Laterality

We conducted a long-term survey on the ratios of laterality in the populations of the two species of scale-eater on a rocky shore at Kasenga Point in the southern end of Lake Tanganyika (near Mpulungu, Northern District, Republic of Zambia). It has been shown that the fish community in the littoral rocky shore of this lake is mainly composed of cichlid fishes and is very stable with the densities of most species being unchanged over many years [17,28,39]. Then, in order to know the effect of laterality of one species on another, it may be enough to know the ratios of laterality of the two species of scale-eater. In the same season (from September to November) in every year, fish samples of the two species were purchased from fish hauls by village fishermen at Kasenga Point. The hauls had been caught as their livelihood mainly by gill-net and sometimes by sein-net and angling every day. In every season, several individuals of each species (ranging from 10 to 200, the average was 132 for *P. microlepis* and 57 for *P. straeleni*) were collected in order to gain reliable ratios of laterality for each population at each time. This survey was conducted almost every year from 1988 to 2018, though we could not make the survey in two years in the beginning of this survey (1989 and 1991).

The laterality of each fish can be defined from the direction of the mouth opening [15,17,21]. Morphologically, the asymmetric mouth opening is due to either side of the joint, say, the right joint, between mandible and suspensorium taking a position frontward, ventrally, and outside compared to the opposite side of the joint [16]. The bending rightwards should mean that the right side of its head and flank have developed more compared to the left side [21]. This relation is seemingly held in other fishes [35,37]. The functional morphology and the quantitative measurement of the asymmetric mouth opening have been developed in separate studies [31,36,40]. Note that the definition of laterality used in the present and recent studies differs from that used in earlier papers [15–17,33]. In the earlier papers, individuals with the mouth opening to the right were defined as "right-handed" or "dextral". In the present and recent studies, usage of "lefty" reflects the fact that the left mandible of such 'right-handed' fish is larger than the right mandible [21,31], and the left eye is dominant [34,41].

Temporal change in the ratios of laterality (frequency of righty morphs in each population) in each species was determined using the samples mentioned above for the whole period, and the periodicity in the ratio of laterality, if any, was analyzed using the Fourier transform with the same method formalized in the Supplementary Material in the work of Yasughi and Hori [36].

2.2. Sampling of Pairs of Hunting Scale-Eaters and Their Victim Fish to Assess the Combination of Laterality at Each Predation Event

This sampling was carried out at Kasenga Point mainly in November of 2007, and supplementary samplings were added at the site in the same season from 2008 to 2018. In every survey, we used a scuba to trace each fish of either species of scale-eater in water 3–10 m deep. When observing the incidence of scale-eating, we immediately tried to catch both the scale-eater and its victim by spreading a short gill-net around them. Then, in the event that both the scale-eater and its victim were caught, the fishes were unloaded, fixed in water with ice, and kept in 10% formol. If only one fish was caught, it was released, and another tracing was started. Success in catching both fishes was achieved about once every 3 or 4 trials. In the laboratory, the laterality of the fishes was determined based on the direction of the mouth opening, and the combinations of their laterality were examined. These treatments were performed according to the Regulations on Animal Experimentation at Kyoto University. To know whether cross-predation or parallel-predation was predominant in each species of scale-eater, a statistical test was performed using odds ratios with a 95% confidential limit.

3. Results

3.1. Temporal Changes in the Laterality of Two Scale-Eaters

In Kasenga point, both lateral morphs of the two species of scale-eater were maintained for 31 years. The ratios of laterality of both species periodically and dynamically changed around a value of 0.5 but were almost always maintained within a range of 0.4 to 0.6 (Figure 2). The Fourier transform detected a significant period of cycle in the ratio of each species, i.e., 3.9 years (p-value < 0.001) for *P. microlepis* and 4.1 years (p-value < 0.01) for *P. straeleni*, indicating that the ratios of the two species oscillated with almost the same period.

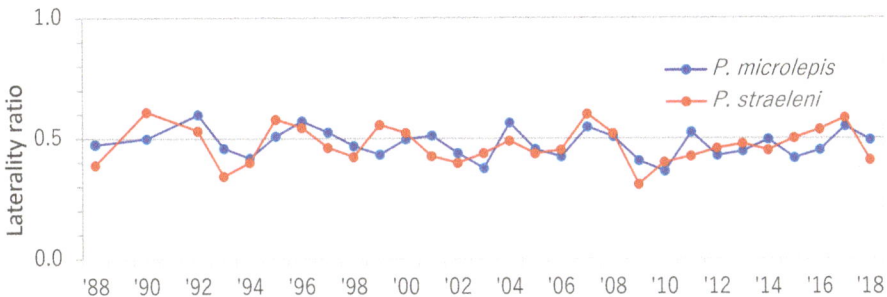

Figure 2. The temporal changes in the ratios of laterality (frequency of righty morphs) in two scale-eaters, *P. microlepis* and *P. straeleni*, on the southernmost shoreline of Lake Tanganyika over a 30-year period from 1988 to 2017.

Relation of oscillation of laterality between the two species of scale-eater was examined by directly plotting the ratio of *P. microlepis* against that of *P. straeleni* (Figure 3). The plots scattered in a counterclockwise rotation around the equilibrium point in both species (coordinates (0.5, 0.5)). It suggests that the ratio of *P. microlepis* followed the periodic change of *P. straeleni* nearly one year later.

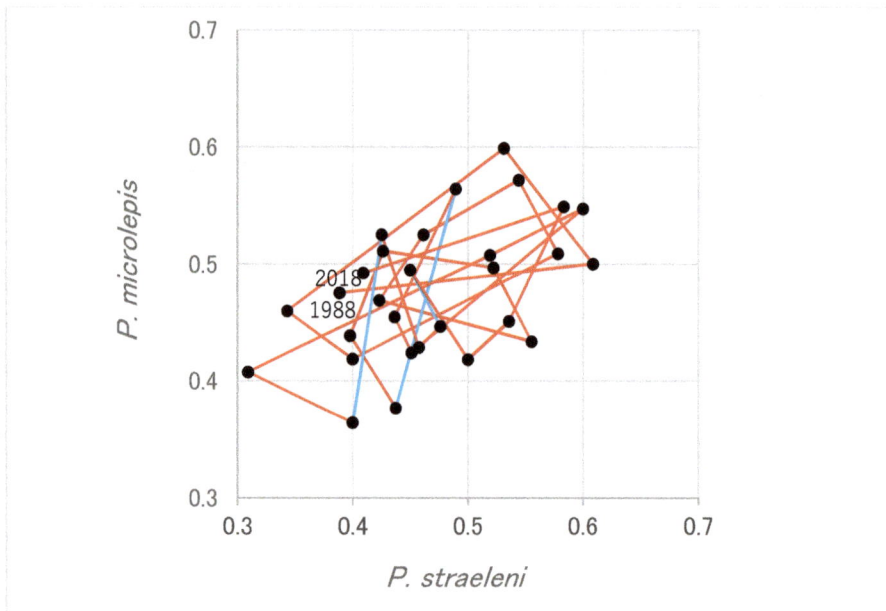

Figure 3. The relationship of laterality ratios between *P. microlepis* and *P. straeleni*. Red lines indicate that the change proceeds anti-clockwise, thus the change in *P. microlepis* follows that in *P. straeleni*. Blue lines proceed clockwise, indicating the opposite pattern.

3.2. Correspondence of Fish Laterality between Individual Scale-Eater and Its Prey

We were able to collect 53 pairs of individual scale-eaters and prey fish for *P. microlepis*, and 51 pairs for *P. straeleni*. The "victim fish" (fish that have their scales eaten) were composed of *Interochromis loocki, Petrochromis* spp., *Tropheus moorii,* and *Lamprologus callipterus,* and they looked to be of little difference to the two species of scale-eater. The combination between the laterality of scale-eater and its prey exhibited a significant bias toward cross-predation in both species (Table 1). The odds ratio was 26.4 for *P. microlepis* (95% confidential limit; 24.9–27.9) and 28.88 for *P. straeleni* (27.4–30.4), indicating that cross-predation significantly occurred more frequently than parallel-predation (Mantel–Haenszel test; $p < 0.01$ for both species).

Table 1. The combination of morph types between each scale-eater and its victim fish. *R* and *L* represent the righty morph and lefty morph, respectively.

Species	Scale-Eater	Victim Fish		Species	Scale-Eater	Victim Fish	
		R	*L*			*R*	*L*
P. microlepis	*R*	3	19	*P. straeleni*	*R*	4	21
	L	25	6		*L*	22	4
		Total: 53				Total: 51	

4. Discussion

4.1. Relation between the Two Species of Scale-Eater

Results showed that the lateral dimorphism of two species of scale-eater cohabiting in a rocky shore was maintained just as it was where only one species (*P. microlepis*) inhabited [15]. The ratios of the two species oscillated during the 30 years with nearly the same period (four years) and the same amplitude (0.2). These patterns indicate that the ratios of laterality of the two species interacted with each other. The change in ratio of *P. microlepis* appeared to follow that of *P. straeleni*.

This semi-synchronized pattern may be interpreted by the frequency-dependent selection (minority advantage) in scale-eating and also by the sharing of advantage between the minor morphs in frequency of the two species. A long-term census carried out in the same site (Takeuchi et al., 2010) [28] indicated that the density of the two species of scale-eater has remained rather constant with that of *P. microlepis* being 8.4 times higher than that of *P. straeleni* (the mean density of *P. microlepis* was 11.73/10 m^2 (S.D., 28.76) and that of *P. straeleni* 1.39 (3.98)). If the minority advantage operates in this system, the minor morph at a time of *P. straeleni*, say, the righty morph, will get the higher fitness, and then the advantage should be shared by the righty morph of *P. microlepis*. This sharing of minority advantage may be the most responsible factor for the ratio of the laterality of majority scale-eaters in number, *P. microlepis*, following the oscillation of the minority one, *P. straeleni*. Using a theoretical model, Takahashi and Hori [26,42] examined the factor that generates an oscillation of polymorphism in two competing predators and found that the strong frequency-dependent selection was crucial for the polymorphism being dynamically maintained. In the next section, we examine the strength of the frequency-dependent selection between scale-eaters and their prey.

4.2. Predator and Prey Relationship

The results showed that righty scale-eaters predominantly succeeded in attacking lefty prey over righty prey, and vice versa for lefty scale-eaters in both *P. microlepis* and *P. straeleni*. A statistical test based on the common odds ratio was highly significant (Mantel–Haenszel test; $p < 0.01$), suggesting that "cross-predation" was 26 to 29 times more frequent than "parallel-predation". Cross-predation should be crucially responsible for the oscillations being generated. Using a theoretical model, Takahashi and Hori [25] examined the factor that could generate an oscillation of polymorphism in a scale-eater and found that the strong frequency-dependent selection was essential for the oscillation around the equilibrium. The high level of predominance of cross-predation found in both scale-eaters seemed to account enough for the oscillation of laterality observed in their populations. In such a situation, the majority morph of scale-eater at any one time, say, righty scale-eaters, would exploit the lefty prey fish. Then the righty prey would be at an advantage in fitness and increase its own frequency after a period of time had elapsed, which would then favor the lefty morph of scale-eater. Furthermore, Nakajima et al. [43], using mathematical analyses and computer simulations, ascertained that under the predominance of cross-predation and high predation efficiency, the dimorphism was dynamically maintained in a one-predator–two-prey system as well as a three-trophic-levels system.

Yasugi and Hori [36] investigated the relationship between largemouth bass, *Micropterus salmoides*, and freshwater gobies, *Rhinogobius* spp., in regards to their laterality, and found a significant bias toward cross-predation. Then, Yasugi and Hori [37] studied experimentally the kinematic causation of predominance of cross-predation. They found that every morph of laterality has a dominant side (either right or left) of the body in sensory and locomotion ability which may function differentially in detecting and attacking a prey and making an attack in predator fish, and detecting and evading an enemy and escaping in prey fish. Then, in the system where a predator stalks and attacks prey fish from behind, predation is more successful when lefty (righty) predator meets a righty (lefty) prey, and less successful when a lefty (righty) predator meets a lefty (righty) prey. Though no quantitative data were taken, we have the impression that scale-eaters can detect the laterality of prey that they targeted at an early stage of their pursuit, because the scale-eaters were often observed to pursue for a long time in cases of cross-predation. We think that they can detect the prey's laterality judging by any delicate gesture in the prey's movement, even though the laterality of each individual seems to be concealed from appearance [21].

Yasugi and Hori [38] also demonstrated that the dominance of cross-predation is characteristic of predators who attack prey from behind. A similar result was found in the investigation of stomach content analysis of the relation between piscivorous fishes belonging to *Lamprologus* (sensu lato) and their prey in regards to their laterality (Hori, unpublish data). Inversely, parallel-predation is dominant in predators who attack the benthic prey fish from ahead such as anglerfish, *Lophinomus setigerus* [38].

In this study we could not analyze the relation between ratios of predator and prey. The littoral fish community in Lake Tanganyika is so speciose and the interspecific interactions are highly complicated [17,27,30]. The interaction between piscivorous fishes and their prey and that between benthos-eating fishes and their prey may also be involved. Therefore, to analyze the temporal relation of laterality between predator and prey in the natural system, a more simple system such as one-predator–one-prey may be suitable. The fish community in a pelagic area in a temperate zone might provide such a system.

Furthermore, cross-predation may have a large effect on the structure of a fish community itself. Using a mathematical model and a computer simulation applied to a fish community of three trophic levels, Nakajima et al. [44] predicted that, when only one type of lateral morph exists in a species, the other type can invade, which suggests that dimorphism is maintained in all directly and indirectly interacting species of a community. Takahashi and Hori [42] theoretically showed that oscillation in laterality of morphs affects the coexistence of competing species. For an investigation of the relation between laterality and community structure in a natural system, however, the fish community in Lake Tanganyika may be suitable in spite of the complexity of its structure. The reason is that, as every population was very stable [17,28], the effect of frequency of each morph of a species on other fishes could be assessed only with periodic fish samplings from the community. This type of research is now in progress at the same site of the present study.

At present, it is difficult to evaluate the effect of frequency-dependent vigilance of prey fish on oscillation. Hori [15] demonstrated that prey fish actually showed such differential vigilance responding to the ambient ratio of laterality in scale-eaters. However, the result was reached in circumstances where the laterality of prey fish had not yet been recognized, and thus the two kinds of effects on oscillation, i.e., cross-predation and differential vigilance, were mingled. As the theoretical model [42] predicts that natural selection should be strong to generate the oscillation of the ratios of laterality, frequency-dependent vigilance of prey fish may have a large effect on the oscillation, as well as the predominance of cross-predation. The evaluation of the relative importance of the vigilance of prey fish remains for future investigation.

Author Contributions: Conceptualization, M.H. and S.T.; methodology, M.H. and S.T.; statistical analysis, S.T.; investigation, M.H., M.K. and S.A.; data collecting, M.H., M.K. and S.A.; data curation, M.H.; writing—original draft preparation, M.H.; writing—review and editing, M.K., S.A., S.T. and M.H.; project administration, M.H.; resources, M.H. and M.K.

Funding: This study was supported by grants from the Ministry of Education, Culture, Science and Technology, Japan (21st Century COE Program (A14), Global Coe Program (A06), Priority Area (14087203), and Scientific Research (B, 21370010, 15H05230, 16H05773, 17K14934) for MH and (B, 16H05773) for MK).

Acknowledgments: We are grateful to members of the Maneno Team (Tanganyika Research Project Team) and staff of LTRU, Mpulungu, Zambia, for their support. The research presented here was conducted with permission for fish research in Lake Tanganyika from the Zambian Ministry of Agriculture, Food, and Fisheries and complied with the current law in Zambia.

Conflicts of Interest: The authors declare no conflict of interest.

References

1. Bisazza, A.; Rogers, L.J.; Vallortigara, G. The origins of cerebral asymmetry: A review of evidence of behavioral and brain lateralization in fishes, reptiles and amphibians. *Neurosci. Biobehav. Rev.* **1998**, *22*, 411–426. [CrossRef]
2. Vallortigara, G.; Rogers, L.J.; Bisazza, A. Possible evolutionary origins of cognitive brain lateralization. *Brain Res. Rev.* **1999**, *30*, 164–175. [CrossRef]
3. Rogers, L.J.; Vallortigara, G.; Andrew, R.J. *Divided Brains: The Biology and Behaviour of Brain Asymmetries*; Cambridge University Press: Cambridge, UK, 2013.

4. Vallortigara, G.; Versace, E. Laterality at the neural, cognitive, and behavioral levels. In *APA Handbooks in Psychology. APA Handbook of Comparative Psychology: Basic Concepts, Methods, Neural Substrate, and Behavior*; Call, J., Burghardt, G.M., Pepperberg, I.M., Snowdon, C.T., Zentall, T., Eds.; American Psychological Association: Washington, DC, USA, 2017; pp. 557–577.

5. MacNeilage, P.; Rogers, L.J.; Vallortigara, G. Evolutionary origins of your left and right brain. *Sci. Am.* **2009**, *301*, 60–67. [CrossRef] [PubMed]

6. Vallortigara, G.; Rogers, L.J. Survival with an asymmetrical brain: Advantages and disadvantages of cerebral lateralization. *Behav. Brain Sci.* **2005**, *28*, 575–633. [CrossRef]

7. Vallortigara, G.; Chiandetti, C.; Sovrano, V.A. Brain asymmetry (animal). *WIREs Cogn. Sci.* **2011**, *2*, 146–157. [CrossRef] [PubMed]

8. Rogers, L.J.; Vallortigara, G. When and why did brains break symmetry? *Symmetry* **2015**, *7*, 2181–2194. [CrossRef]

9. Palmer, A.R.; Strobeck, C. Fluctuating asymmetry as a measure of development stability: Implications of non-normal distributions and power of statistical tests. *Acta Zool. Fennica* **1992**, *191*, 57–72.

10. Rogers, L.J.; Andrew, R.J. *Comparative Vertebrate Lateralization*; Cambridge University Press: London, UK, 2002; ISBN 0-521-78161-2.

11. Ströckens, F.; Gunturkün, O.; Ocklenburg, S. Limb preferences in non-human vertebrates. *Laterality Asymmetries Body Brain Cogn.* **2013**, *18*, 536–575. [CrossRef]

12. Ghirlanda, S.; Vallortigara, G. The evolution of brain lateralization: A game-theoretical analysis of population structure. *Proc. R. Soc. Lond. B* **2004**, *271*, 853–857. [CrossRef] [PubMed]

13. Ghirlanda, S.; Frasnelli, E.; Vallortigara, G. Intraspecific competition and coordination in the evolution of lateralization. *Philos. Tranzactions R. Soc. B* **2009**, *364*, 861–866. [CrossRef] [PubMed]

14. Vallortigara, G. The Evolutionary psychology of left and right: Costs and benefits of lateralization. *Dev. Phychobiol.* **2006**, *48*, 418–427. [CrossRef] [PubMed]

15. Hori, M. Frequency-dependent natural selection in the handedness of scale-eating cichlid fish. *Science* **1993**, *260*, 216–219. [CrossRef] [PubMed]

16. Liem, K.F.; Stewart, D.J. Evolution of the scale-eating cichlid fishes of Lake Tanganyika: A generic revision with a description of a new species. *Bull. Mus. Comp. Zool.* **1976**, *147*, 319–350.

17. Hori, M. Feeding relationships among cichlid fishes in Lake Tanganyika: Effects of intra- and interspecific variation of feeding behavior on their coexistence. *Ecol. Int. Bull.* **1991**, *19*, 89–101.

18. Takahashi, R.; Moriwaki, T.; Hori, M. Foraging behaviour and functional morphology of two scale-eating cichlids from Lake Tanganyika. *J. Fish Biol.* **2007**, *70*, 1458–1469. [CrossRef]

19. Takeuchi, Y.; Hori, M.; Oda, Y. Lateralized kinematics of predation behavior in a Lake Tanganyika scale-eating cichlid fish. *PLoS ONE* **2012**, *7*, e29272. [CrossRef] [PubMed]

20. Hata, H.; Yasugi, M.; Hori, M. Jaw laterality and related handedness in the hunting behavior of a scale-eating characin, *Exodon paradoxus*. *PLoS ONE* **2011**, *6*, e29349. [CrossRef]

21. Hori, M.; Ochi, H.; Kohda, M. Inheritance pattern of lateral dimorphism in two cichlids (a scale eater, *Perissodus microlepis*, and an herbivore, *Neolamprologus moorii*) in Lake Tanganyika. *Zool. Sci.* **2007**, *24*, 489–492. [CrossRef] [PubMed]

22. Stewart, T.A.; Albertson, R.C. Evolution of a unique predatory feeding apparatus: Functional anatomy, development and a genetic locus for jaw laterality in Lake Tanganyika scale-eating cichlids. *BMC Biol.* **2010**, *8*, 8. [CrossRef] [PubMed]

23. Hata, H.; Takahashi, R.; Ashiwa, H.; Awata, S.; Takeyama, T.; Kohda, M.; Hori, M. Inheritance patterns of lateral dimorphism examined by breeding experiments with the Tanganyikan cichlid (*Julidochromis transcriptus*) and the Japanese medaka (*Oryzias latipes*). *Zool. Sci.* **2012**, *29*, 49–53. [CrossRef]

24. Hata, H.; Hori, M. Inheritance patterns of morphological laterality in mouth opening of zebrafish, *Danio rerio*. *Laterality* **2012**, *17*, 741–754. [CrossRef] [PubMed]

25. Takahashi, S.; Hori, M. Unstable evolutionarily stable strategy and oscillation: A model on lateral asymmetry in scale-eating cichlids. *Am. Nat.* **1994**, *144*, 1001–1020. [CrossRef]

26. Takahashi, S.; Hori, M. Oscillation maintains polymorphisms—A model of lateral asymmetry in two competing scale-eating cichlids. *J. Theor. Biol.* **1998**, *195*, 1–12. [CrossRef]

27. Hori, M. Structure of littoral fish communities organized by their feeding activities. In *Fish Communities in Lake Tanganyika*; Kawanabe, H., Hori, M., Nagoshi, M., Eds.; Kyoto University Press: Kyoto, Japan, 1997; pp. 275–298. ISBN 4-87698-042-X.

28. Takeuchi, Y.; Ochi, H.; Kohda, M.; Sinyinza, D.; Hori, M. A 20-year census of a rocky littoral fish community in Lake Tanganyika. *Ecol. Freshw. Fish* **2010**, *19*, 239–248. [CrossRef]

29. Takahashi, R.; Watanabe, K.; Nishida, M.; Hori, M. Evolution of feeding secialization in Tanganyikan scale-eating cichlids: A molecular phylogenetic approach. *BMC Evol.* **2007**, *7*, 195.

30. Hori, M. Mutualism and commensalism in the fish community of Lake Tanganyika. In *Evolution and Coadaptation in Biotic Communities*; Kawano, S., Connell, J.H., Hidaka, T., Eds.; Tokyo University Press: Tokyo, Japan, 1987; pp. 219–239.

31. Hori, M.; Nakajima, M.; Hata, H.; Yasugi, M.; Takahashi, S.; Nakae, M.; Yamaoka, K.; Kohda, M.; Kitamura, J.; Maehata, M.; et al. Laterality is universal among fishes but increasingly cryptic among derived groups. *Zool. Sci.* **2017**, *34*, 1–8. [CrossRef] [PubMed]

32. Nakajima, M.; Yodo, T.; Katano, O. Righty fish are hooked on the right side of their mouths—Observations from an angling experiment with largemouth bass, *Microperus salmoides*. *Zool. Sci.* **2007**, *24*, 855–859. [CrossRef]

33. Seki, S.; Kohda, M.; Hori, M. Asymmetry of mouth morph of a freshwater goby, *Rhinogobius flumineus*. *Zool. Sci.* **2000**, *17*, 1321–1325. [CrossRef]

34. Matsui, S.; Takeuchi, Y.; Hori, M. Relation between morphological antisymmetry and behavioral laterality in a Poeciliid fish. *Zool. Sci.* **2013**, *30*, 613–618. [CrossRef]

35. Takeuchi, Y.; Hori, M. Behavioural laterality in the shrimp-eating cichlid fish, *Neolamprologus fasciatus*, in Lake Tanganyika. *Anim. Behav.* **2008**, *75*, 1359–1366. [CrossRef]

36. Yasugi, M.; Hori, M. Predominance of cross-predation between lateral morphs in a largemouth bass and a freshwater goby. *Zool. Sci.* **2011**, *28*, 869–874. [CrossRef] [PubMed]

37. Yasugi, M.; Hori, M. Lateralized behavior in the attacks of largemouth bass on *Rhinogobius* gobies corresponding to their morphological antisymmetry. *J. Exp. Biol.* **2012**, *215*, 2390–2398. [CrossRef] [PubMed]

38. Yasugi, M.; Hori, M. Predominance of parallel- and cross-predation in anglerfish. *Mar. Ecol.* **2016**, *37*, 576–587. [CrossRef]

39. Hori, M.; Yamaoka, K.; Takamura, K. Abundance and micro-distribution of cichlid fishes on a rocky shore of Lake Tanganyika. *Afr. Study Monogr.* **1983**, *3*, 25–38.

40. Hata, H.; Yasugi, M.; Takeuchi, Y.; Hori, M. Distinct lateral dimorphism in the jaw morphology of the scale-eating cichlids, *Perissodus microlepis* and *P. straeleni*. *Ecol. Evol.* **2013**, *3*, 4641–4647. [CrossRef] [PubMed]

41. Takeuchi, Y.; Hori, M.; Myint, O.; Kohda, M. Lateral bias of agonistic responses to mirror images and morphological asymmetry in the Siamese fighting fish (*Betta splendens*). *Behav. Brain Res.* **2010**, *208*, 106–111. [CrossRef] [PubMed]

42. Takahashi, S.; Hori, M. Coexistence of competing species by the oscillation polymorphism. *J. Theor. Biol.* **2005**, *235*, 591–596. [CrossRef]

43. Nakajima, M.; Matsuda, H.; Hori, M. A population genetic model for lateral dimorphism frequency in fishes. *Am. Nat.* **2005**, *163*, 692–698. [CrossRef]

44. Nakajima, M.; Matsuda, H.; Hori, M. Persistence and fluctuation of lateral dimorphism in fishes. *Am. Nat.* **2004**, *163*, 692–698. [CrossRef]

symmetry

MDPI

Article

Incubation under Climate Warming Affects Behavioral Lateralisation in Port Jackson Sharks

Catarina Vila Pouca *, Connor Gervais, Joshua Reed and Culum Brown

Department of Biological Sciences, Macquarie University, Sydney 2109, Australia;
connor-robert.gervais@students.mq.edu.au (C.G.); joshua.reed@students.mq.edu.au (J.R.);
culum.brown@mq.edu.au (C.B.)
* Correspondence: catarina.vilapouca@mq.edu.au

Received: 10 May 2018; Accepted: 25 May 2018; Published: 28 May 2018

Abstract: Climate change is warming the world's oceans at an unprecedented rate. Under predicted end-of-century temperatures, many teleosts show impaired development and altered critical behaviors, including behavioral lateralisation. Since laterality is an expression of brain functional asymmetries, changes in the strength and direction of lateralisation suggest that rapid climate warming might impact brain development and function. However, despite the implications for cognitive functions, the potential effects of elevated temperature in lateralisation of elasmobranch fishes are unknown. We incubated and reared Port Jackson sharks at current and projected end-of-century temperatures and measured preferential detour responses to left or right. Sharks incubated at elevated temperature showed stronger absolute laterality and were significantly biased towards the right relative to sharks reared at current temperature. We propose that animals reared under elevated temperatures might have more strongly lateralized brains to cope with deleterious effects of climate change on brain development and growth. However, far more research in elasmobranch lateralisation is needed before the significance of these results can be fully comprehended. This study provides further evidence that elasmobranchs are susceptible to the effects of future ocean warming, though behavioral mechanisms might allow animals to compensate for some of the challenges imposed by climate change.

Keywords: laterality; climate change; temperature; development; elasmobranchs

---

1. Introduction

Climate change has been identified as one of the major human-induced environmental impacts to ecosystems worldwide [1]. The average temperature of the upper layers of the ocean has increased by 1.0 °C over the past 120 years, and is predicted to increase by 1–3 °C over the next century if the current trajectory of greenhouse gas emissions is maintained [1,2]. In addition, oceanic carbon dioxide (CO_2) levels have now reached historically high levels [3]. Such rapid changes in important environmental parameters will considerably impact marine ecosystems.

Elevated temperatures and CO_2 levels in the ocean can significantly impair sensory functions and alter critical behavior in teleost fish and elasmobranchs. For example, coral reef fish and benthic sharks exposed to elevated CO_2 levels showed impaired olfactory and auditory responses, important for predator/prey recognition and homing behavior [4–9]. Additionally, exposure to elevated temperatures resulted in higher developmental rate and metabolism, as well as limited growth, aerobic scope, reproductive output, and foraging [8–12]. Whereas highly mobile species will likely shift their distributions poleward [13], less mobile species will have to cope with these changes through rapid evolution or phenotypic plasticity. Ectotherms are especially vulnerable to global warming because their body temperature and basic physiological functions are regulated by the external environment. In addition, many elasmobranch species are oviparous and have long gestation periods of several months, so embryos will be exposed to prevailing environmental conditions and have little choice other than to adapt or die. One mechanism used by more sedentary species to compensate for the increase in developmental and metabolic rates is the reallocation of energy resources during development, which is expected to affect highly metabolically expensive systems, such as neural development and processing [14–16]. The detrimental effects in a range of sensory modalities and behaviors have already been observed in fish (e.g., [4,6]), suggesting that predicted climate change conditions might disproportionately impact brain development and function.

Behavioral lateralisation, the tendency to favor the left or right side in a given context, results from a functional asymmetry between the two hemispheres of the brain [17–20]. Cerebral and behavioral lateralisations are widespread in the animal kingdom and have been well studied in teleost fish [17,21]. Teleosts generally prefer to use the left eye and right hemisphere to process biologically relevant stimuli, such as predators or potential mates, and emotional responses, such as fear and aggression, whereas the right eye and left hemisphere are generally linked to stimuli categorisation and object manipulation. Nonetheless, we often see species, population, or individual differences that arise through a mixture of genetic and experiential effects [22–25].

Laterality in elasmobranchs is understudied, with only two studies investigating behavioral lateralisation in benthic sharks [26,27]. Byrnes, Vila Pouca and Brown [26] observed individual levels of laterality bias in rotational swimming and T-maze turn preference in juvenile Port Jackson sharks, with females more strongly lateralised than males, and Green and Jutfelt [27] reported a population-level laterality bias in double T-maze turn in small-spotted catsharks. Lateralisation of behavior and cognitive functions have been suggested to offer selective advantages [19,28]. For example, laterality enhances schooling behavior that can have important fitness-related implications in foraging and anti-predator behavior [29]. Schools of lateralised fish were more cohesive and coordinated than schools of non-lateralised fish [30], and individuals tended to take positions in the school that correspond to their visual hemifield preferences for observing conspecifics [25]. A laterality bias might also provide them with advantages in multitasking situations, such as foraging while being vigilant to predators, and enhancing cognitive efficiency [28,31–33].

Since behavioral lateralisation is an expression of brain function, it can be used as a barometer of normal brain development and function in some contexts, namely exposure or development under climate change conditions. An increasing number of studies have reported the impact of increased CO_2 levels and elevated temperature on behavioral lateralisation in some teleost species, though with varying direction and magnitude [34–38]. The behavioral effects of elevated CO_2 levels in teleosts seem to stem from a dysfunction of the GABA-A neurotransmitter receptor in the brain [39]. In elasmobranchs, only one study investigated the effects of future climate change conditions on behavioral lateralisation [27]. Small-spotted catsharks aged 4–24 months exposed to increased CO_2 for as little as four hours showed stronger absolute lateralisation at the population level when compared with control individuals [27]. Such short-term responses are likely indicative of phenotypic plasticity and might mimic responses to brief environmental changes (e.g., day vs. night or intertidal zone conditions). The impacts of long-term exposure to elevated temperature on cerebral lateralisation, especially during critical developmental periods, have not yet been assessed in elasmobranchs. With so many reported consequences on development and physiology in elasmobranchs [8,9,40,41], it is likely that rapid climate warming might also impact brain function in this group. In the present study, we tested the hypothesis that the predicted end-of-century temperature during embryogenic and hatchling development affects behavioral lateralization in benthic shark species.

2. Materials and Methods

2.1. Ethics Statement

Egg collection occurred under New South Wales Fisheries permit P08/0010-4.2. The experiments were approved by the Macquarie University Animal Ethics Committee (ARA 2016-027). All animals were euthanised at the end of the experiment with a lethal dose of MS-222 (tricaine methane-sulfonate; 1.5 g/L seawater) for brain anatomy studies, to be reported in a separate paper.

2.2. Egg Collection and Incubation

We collected Port Jackson shark eggs via snorkeling from Jervis Bay, NSW. Females lay their eggs in shallow rocky reefs during late winter, mostly during August and September [42]. Freshly laid egg capsules are clean, soft, pliable, and olive green in color, but become brittle after two weeks and change to a dark brown color in 3–5 weeks [43] enabling an estimation of laying date. Under ambient conditions, embryos have a long incubation period of 10–11 months [43]. Eggs were collected on 11 October and 2 November 2016 and we estimate all had been laid recently, within 6 weeks of collection. Eggs were transported to Macquarie University, Sydney Australia, and held in a temperature-controlled laboratory until hatching. The eggs were placed in 40 L tanks containing natural filtered seawater and the temperature was maintained using a custom-design Seawater Environmental Control Mixing Chamber. Following transport, eggs were left to rest for 7 days, then the temperature was steadily increased by 0.5 °C/day to the elevated temperature treatment in half of the tanks. We randomly divided eggs among the two treatments: a control temperature treatment (C; $n = 12$) incubated at 20.6 ± 0.5 °C, consistent with the annual average maximum temperature in Jervis Bay; and an elevated temperature treatment (ET; $n = 12$) incubated at 23.6 ± 0.5 °C, representing the end of century projected sea-surface temperature increase under the representative concentration pathway (RCP) 8.5 climate model [1]. The CO_2 of the system reflected ambient conditions (ρCO_2 ~418 ppm).

When the egg capsules' mucous plugs opened, approximately four months into development, the embryos were removed from the egg and placed in individual containers within the housing tank for close monitoring.

2.3. Husbandry and Rearing

Approximately one month after hatching (stage 15 [43]; external yolk completely exhausted, internal yolk virtually depleted, and disappearance of slime coat), individuals were moved to the Sydney Institute of Marine Science (SIMS). Sharks were housed in groups of six animals in 1000 L tanks maintained at incubation temperatures using submersible heaters (one 2000 W titanium stick heater or four 300 W AquaOne glass heaters). Tanks had a continuous supply of fresh seawater pumped directly from Sydney harbor, aeration, and polyvinyl chloride (PVC) structures and fake kelp to provide shelter and enrichment. Tanks were scrubbed clean at least once a week. The room had a natural light/dark cycle.

Immediately after hatching, Port Jackson sharks were weighed, measured (total length, TL), and individually tagged beneath the dorsal fin (Passive Integrated Nano-Transponder, Trovan® ID-100A/1.25, Microchips Australia Pty Ltd., Victoria, Australia). Sharks were fed ad libitum on a mixed diet of squid, cuttlefish, whitebait, and prawns three times per week.

Five sharks from the elevated temperature treatment did not survive the first month after hatching (three deaths and two were euthanised because they were not feeding). We therefore started the procedure with seven ET and twelve C sharks, 58.3 and 100% of our initial sample size for each group, respectively. The median age of the test subjects from the ET group was 63.5 days post-hatching compared with 85.5 of the C group.

2.4. Procedure

The experimental tank (180 × 100 × 40 cm; Figure 1) was maintained at incubation temperature using four to six 300 W AquaOne glass heaters. For four days prior to the laterality assay, sharks were allowed to familiarise with the experimental tank to allow them to overcome any stress associated with moving between the housing and experimental tanks and being in a novel environment. During the familiarisation phase, the shark could swim freely in the tank for a 30-min period after which it was fed 2% of its body weight.

To assess behavioral laterality, sharks were tested individually in a detour test using a Y-maze [44]. The test consisted of 10 trials conducted on a single day. For each trial, the subject was ushered down a corridor and its turn direction at the end of the maze was recorded. Based on results from a pilot study, a small piece of food was placed behind the partition at the end of the corridor and individuals were fed at the end of the 5th and 10th trials to encourage directed swimming along the maze and ensure motivation in the task. After each run, the shark was temporarily constrained in the choice zone. The subject was then released and allowed to swim down the runway in the opposite direction. This approach reduces handling stress and reduces the possibility of extraneous cues inducing side biases. The shark was allowed 30 s to recover between runs.

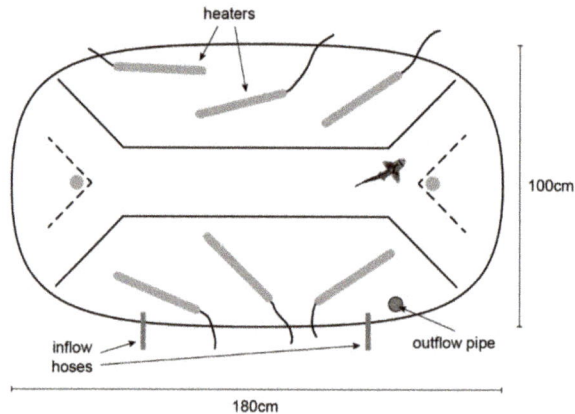

Figure 1. Diagram of the experimental tank.

2.5. Data Analysis

We calculated laterality index as follows: L_I = (number of right turns − number of left turns)/(total number of turns). L_I is a continuous value from −1 to 1, in which a positive score indicated a preference for rightward turns and a negative score indicated a preference for leftward turns. Laterality strength (L_S) was calculated by taking the absolute value of L_I.

Statistical analyses were conducted in R v. 3.4.3 [45]. We used non-parametric tests due to low sample sizes. Separate Mann–Whitney U tests were used to compare L_I and L_S scores between C and ET individuals, and to test if sharks within each group were significantly lateralised (distribution with $\mu \neq 0$).

3. Results

Sharks from the elevated temperature treatment (ET) showed stronger absolute laterality (L_S) compared to control temperature (C) sharks (Figure 2a; $W = 19$, $P = 0.047$), along with higher laterality index (L_I) values (Figure 2b; $W = 10.5$, $P = 0.0067$). ET sharks displayed a significant rightward bias ($V = 28$, $P = 0.021$), whereas C sharks showed no population-level preference for either side ($V = 12.5$, $P = 0.746$). Individual turn preferences are provided in Table 1.

We examined the possible effect of age within the control group and found no relationship between age at testing and L_I (d.f. = 10, $t = -0.06$, $P = 0.953$, $R^2 = 0.00036$).

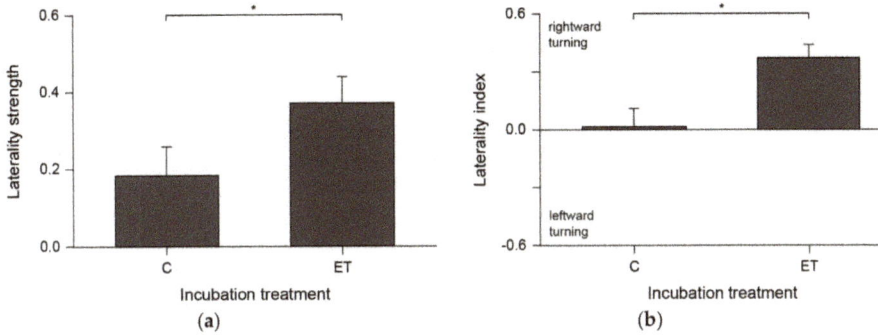

Figure 2. (a) Laterality strength (group mean ± SEM) and (b) laterality index (group mean ± SEM) in sharks incubated at control temperature (C; $n = 12$) or elevated temperature (ET; $n = 7$).

Table 1. Summary information on experimental subjects and individual left or right turn preference in the detour task.

Shark ID	Sex	Weight (g)	Treatment	# Right Turns	# Left Turns	L_I	L_S
C489	M	86	C	1	9	−0.8	0.8
C451	M	87	C	4	6	−0.2	0.2
C430	M	81	C	5	5	0	0
C437	M	70	C	5	5	0	0
C456	F	53	C	5	5	0	0
C469	M	101	C	5	5	0	0
C492	F	79	C	5	5	0	0
C500	F	89	C	5	5	0	0
C407	M	83	C	6	4	0.2	0.2
C452	F	76	C	6	4	0.2	0.2
C459	M	94	C	6	4	0.2	0.2
C460	F	95	C	8	2	0.6	0.6
ET455	M	64	ET	6	4	0.2	0.2
ET369	M	50.5	ET	6	4	0.2	0.2
ET373	F	59	ET	6	4	0.2	0.2
ET480	M	64.5	ET	7	3	0.4	0.4
ET400	F	78.5	ET	7	3	0.4	0.4
ET433	F	79	ET	8	2	0.6	0.6
ET468	F	62	ET	8	2	0.6	0.6

Note: M, male; F, female; C, control temperature; ET, elevated temperature; L_I, Laterality index; L_S, Laterality strength.

4. Discussion

In this study, we showed that incubation the temperatures predicted for the end of the century affect behavioral lateralisation in Port Jackson sharks. This is the first documented case of a change in lateralised behavior induced by elevated temperature in any elasmobranch. Our hatchling sharks that were incubated and reared in elevated temperature showed stronger absolute laterality and a rightward bias compared with control individuals. Byrnes, Vila Pouca and Brown [26] observed high individual variation in laterality in wild-caught juvenile Port Jackson sharks similar to our control group, suggesting the results from our sharks reared at current ocean temperature in captivity reflect those in wild populations and were not influenced by captive rearing.

The shift in laterality to the right observed in the present study was not clearly due to plastic responses during development or the deaths of left biased or non-lateralised sharks during early ontogeny (42% of sharks reared in elevated temperatures died prior to testing). Behavioral lateralization, in particular handedness, is linked to immune response in humans, rodents, and dogs [46–48]. Immune responses might possibly differ between our two groups. However, to the best of our knowledge, the link between immune competency and lateralisation had not yet been investigated in teleosts or elasmobranchs. Elevated temperature significantly increased developmental rates and metabolism [8,9], with associated costs in terms of energy allocation to growth and physiological processes (e.g., [40]). Therefore, stronger lateralisation may arise as an energy saving mechanism. Functional asymmetries in the brain are thought to enable separate and parallel information processing in each hemisphere, which might increase the brain's capacity to perform simultaneous processing resulting in enhanced cognitive efficiency [21,28]. Neural processing is metabolically expensive; thus, higher parallel processing abilities could allow animals to save energy during brain development and information processing without compromising function. We therefore predict that animals reared under elevated temperatures might have smaller but more strongly lateralized brains. Interestingly, juvenile small-spotted catsharks exposed short-term to elevated CO_2 levels showed stronger absolute laterality in a detour task [27]. Laterality can vary with age [49,50], but we examined the possible effect of age within the control group and found no correlation. It is worth noting that the variation in age within the control group was 35 days, which covers the average age difference between the control and elevated temperature treatments. Future research is needed to determine if laterality varies with age in sharks, perhaps over larger time frames. Regardless of the mechanism, it is apparent that climate change will impact elasmobranchs and the early developmental stages are particularly vulnerable, so further work is required specifically in the context of brain development and cognition under future climate scenarios.

With so few studies investigating laterality in elasmobranchs, commenting on the expected variability at the population or individual level is difficult, let alone on context-specific individual variation. Teleost fish show high variability in laterality strength and direction at the individual, population, and species level [23,44,51]. Additionally, laterality in teleosts has been linked to life history traits and environmental variables [52,53]. Fish from high predation areas, for example, showed stronger laterality than those from low predation areas and this has been linked to schooling behavior in several species [23,25,30,51]. To further complicate the situation, exposure to elevated temperature or CO_2 levels resulted in varying directions and magnitude of change in laterality in different teleosts [34–38]. Some of these different effects might be due to the context of the task or a consequence of altered emotional states of the animal. For example, Domenici, Allan, Watson, McCormick and Munday [35] observed a reversal from right-turning bias in damselfish detouring around an opaque barrier to a left-turning bias when exposed to elevated CO_2 levels. The authors suggested that elevated stress could induce this shift since stressed animals predominantly use the right hemisphere to control motor functions [35,54]. This was also possibly true in the present study, but we assumed that Port Jackson sharks predominantly use the left hemisphere to control motor function when under stress. Further studies are required to determine if this is the case.

5. Conclusions

To conclude, this study provides strong evidence that predicted end-of-century temperature affects behavioural lateralisation in sharks. The combination of elevated CO_2 and temperature might have synergistic effects on laterality. We propose that enhanced lateralisation could help animals cope with the deleterious effects of climate change on development and growth. Evidently, far more research is needed in multiple elasmobranch species to provide an adequate picture of brain lateralisation in elasmobranchs to test this hypothesis. Future studies should combine laterality and cognitive tasks to assess if cognitive functions in elasmobranchs are lateralised, and evaluate the effect of climate change conditions on cognitive performance.

Author Contributions: Conceptualisation: C.V.P., C.G. and C.B.; Methodology: C.G., C.V.P. and C.B.; Investigation: C.G., C.V.P. and J.R.; Analysis: C.V.P.; Writing: C.V.P. and C.B.

Acknowledgments: This research was funded by the Department of Biological Sciences at Macquarie University, and C.V.P. was supported by an Endeavour Postgraduate (PhD) Scholarship. We thank the members and interns of The Fish Lab and staff at SIMS, in particular Andrew Niccum, for husbandry and aquarium maintenance assistance.

Symmetry **2018**, *10*, 184

Conflicts of Interest: The authors declare no conflict of interest.

References

1. Collins, M.; Knutti, R.; Arblaster, J.; Dufresne, J.-L.; Fichefet, T.; Friedlingstein, P.; Gao, X.; Gutowski, W.; Johns, T.; Krinner, G. Long-term climate change: Projections, commitments and irreversibility. In *Climate Change 2013: The Physical Science Basis*; IPCC Working Group I Contribution to AR5; Cambridge University Press: Cambridge, UK; New York, NY, USA, 2013; pp. 1029–1136.

2. Pörtner, H.-O.; Karl, D.M.; Boyd, P.W.; Cheung, W.; Lluch-Cota, S.E.; Nojiri, Y.; Schmidt, D.N.; Zavialov, P.O.; Alheit, J.; Aristegui, J. Ocean systems. In *Climate Change 2014: Impacts, Adaptation, and Vulnerability. Part A: Global and Sectoral Aspects*; Contribution of Working Group II to the Fifth Assessment Report of the Intergovernmental Panel on Climate Change; Cambridge University Press: Cambridge, UK; New York, NY, USA, 2014; pp. 411–484.

3. Stocker, T.F.; Qin, D.; Plattner, G.-K.; Alexander, L.V.; Allen, S.K.; Bindoff, N.L.; Bréon, F.-M.; Church, J.A.; Cubasch, U.; Emori, S. *Climate Change 2013: The Physical Science Basis*; Contribution of Working Group I to the Fifth Assessment Report of the Intergovernmental Panel on Climate Change; Cambridge University Press: Cambridge, UK; New York, NY, USA, 2013; pp. 33–115.

4. Cripps, I.L.; Munday, P.L.; McCormick, M.I. Ocean acidification affects prey detection by a predatory reef fish. *PLoS ONE* **2011**, *6*, e22736. [CrossRef] [PubMed]

5. Munday, P.L.; Cheal, A.J.; Dixson, D.L.; Rummer, J.L.; Fabricius, K.E. Behavioural impairment in reef fishes caused by ocean acidification at CO_2 seeps. *Nat. Clim. Chang.* **2014**, *4*, 487. [CrossRef]

6. Simpson, S.D.; Munday, P.L.; Wittenrich, M.L.; Manassa, R.; Dixson, D.L.; Gagliano, M.; Yan, H.Y. Ocean acidification erodes crucial auditory behaviour in a marine fish. *Biol. Lett.* **2011**, *7*, 917–920. [CrossRef] [PubMed]

7. Dixson, D.L.; Munday, P.L.; Jones, G.P. Ocean acidification disrupts the innate ability of fish to detect predator olfactory cues. *Ecol. Lett.* **2010**, *13*, 68–75. [CrossRef] [PubMed]

8. Rosa, R.; Baptista, M.; Lopes, V.M.; Pegado, M.R.; Ricardo Paula, J.; Trübenbach, K.; Leal, M.C.; Calado, R.; Repolho, T. Early-life exposure to climate change impairs tropical shark survival. *Proc. R. Soc. B Biol. Sci.* **2014**, *281*. [CrossRef] [PubMed]

9. Pistevos, J.C.; Nagelkerken, I.; Rossi, T.; Olmos, M.; Connell, S.D. Ocean acidification and global warming impair shark hunting behaviour and growth. *Sci. Rep.* **2015**, *5*, 16293. [CrossRef] [PubMed]

10. Donelson, J.M.; Munday, P.L.; McCormick, M.I.; Pankhurst, N.W.; Pankhurst, P.M. Effects of elevated water temperature and food availability on the reproductive performance of a coral reef fish. *Mar. Ecol. Prog. Ser.* **2010**, *401*, 233–243. [CrossRef]

11. Munday, P.L.; Kingsford, M.J.; O'Callaghan, M.; Donelson, J.M. Elevated temperature restricts growth potential of the coral reef fish *acanthochromis polyacanthus*. *Coral Reefs* **2008**, *27*, 927–931. [CrossRef]

12. Nilsson, G.E.; Crawley, N.; Lunde, I.G.; Munday, P.L. Elevated temperature reduces the respiratory scope of coral reef fishes. *Glob. Chang. Biol.* **2009**, *15*, 1405–1412. [CrossRef]

13. Perry, A.L.; Low, P.J.; Ellis, J.R.; Reynolds, J.D. Climate change and distribution shifts in marine fishes. *Science* **2005**, *308*, 1912–1915. [CrossRef] [PubMed]

14. Soengas, J.L.; Aldegunde, M. Energy metabolism of fish brain. *Comp. Biochem. Physiol. Part B Biochem. Mol. Biol.* **2002**, *131*, 271–296. [CrossRef]

15. Brown, C. Experience and learning in changing environments. In *Behavioural Responses to a Changing World: Mechanisms and Consequences*; Candolin, U., Wong, B.B., Eds.; Oxford University Press: Oxford, UK, 2012.

16. Sheridan, J.A.; Bickford, D. Shrinking body size as an ecological response to climate change. *Nat. Clim. Chang.* **2011**, *1*, 401. [CrossRef]

17. Rogers, L.J.; Vallortigara, G.; Andrew, R.J. *Divided Brains: The Biology and Behaviour of Brain Asymmetries*; Cambridge University Press: Cambridge, UK, 2013.

18. Vallortigara, G.; Versace, E. Laterality at the neural, cognitive, and behavioral levels. In *Apa Handbook of Comparative Psychology: Basic Concepts, Methods, Neural Substrate, and Behavior, Vol. 1*; American Psychological Association: Washington, DC, USA, 2017; pp. 557–577.

19. Vallortigara, G.; Rogers, L.J. Survival with an asymmetrical brain: Advantages and disadvantages of cerebral lateralization. *Behav. Brain Sci.* **2005**, *28*, 575–588. [CrossRef] [PubMed]

20. Rogers, L.J.; Andrew, R. *Comparative Vertebrate Lateralization*; Cambridge University Press: Cambridge, UK, 2002.
21. Bisazza, A.; Brown, C. Lateralization of cognitive functions in fish. In *Fish Cognition and Behavior*; Wiley: Oxford, UK, 2011; pp. 298–324.
22. Dadda, M.; Bisazza, A. Lateralized female topminnows can forage and attend to a harassing male simultaneously. *Behav. Ecol.* **2006**, *17*, 358–363. [CrossRef]
23. Bisazza, A.; Cantalupo, C.; Capocchiano, M.; Vallortigara, G. Population lateralisation and social behaviour: A study with 16 species of fish. *Later. Asymmet. Body Brain Cogn.* **2000**, *5*, 269–284. [CrossRef] [PubMed]
24. Bisazza, A.; de Santi, A. Lateralization of aggression in fish. *Behav. Brain Res.* **2003**, *141*, 131–136. [CrossRef]
25. Bibost, A.-L.; Brown, C. Laterality influences schooling position in rainbowfish, *melanotaenia* spp. *PLoS ONE* **2013**, *8*, e80907. [CrossRef] [PubMed]
26. Byrnes, E.E.; Vila Pouca, C.; Brown, C. Laterality strength is linked to stress reactivity in port jackson sharks (*heterodontus portusjacksoni*). *Behav. Brain Res.* **2016**, *305*, 239–246. [CrossRef] [PubMed]
27. Green, L.; Jutfelt, F. Elevated carbon dioxide alters the plasma composition and behaviour of a shark. *Biol. Lett.* **2014**, *10*, 20140538. [CrossRef] [PubMed]
28. Rogers, L.J.; Zucca, P.; Vallortigara, G. Advantages of having a lateralized brain. *Proc. R. Soc. Lond. B Biol. Sci.* **2004**, *271*, S420–S422. [CrossRef] [PubMed]
29. Krause, J.; Hoare, D.; Krause, S.; Hemelrijk, C.K.; Rubenstein, D.I. Leadership in fish shoals. *Fish Fish.* **2000**, *1*, 82–89. [CrossRef]
30. Bisazza, A.; Dadda, M. Enhanced schooling performance in lateralized fishes. *Proc. R. Soc. Lond. B Biol. Sci.* **2005**, *272*, 1677–1681. [CrossRef] [PubMed]
31. Dadda, M.; Bisazza, A. Does brain asymmetry allow efficient performance of simultaneous tasks? *Anim. Behav.* **2006**, *72*, 523–529. [CrossRef]
32. Sovrano, V.A.; Dadda, M.; Bisazza, A. Lateralized fish perform better than nonlateralized fish in spatial reorientation tasks. *Behav. Brain Res.* **2005**, *163*, 122–127. [CrossRef] [PubMed]
33. Bibost, A.L.; Brown, C. Laterality influences cognitive performance in rainbowfish melanotaenia duboulayi. *Anim. Cogn.* **2014**, *17*, 1045–1051. [CrossRef] [PubMed]
34. Domenici, P.; Allan, B.; McCormick, M.I.; Munday, P.L. Elevated carbon dioxide affects behavioural lateralization in a coral reef fish. *Biol. Lett.* **2011**. [CrossRef] [PubMed]
35. Domenici, P.; Allan, B.J.; Watson, S.-A.; McCormick, M.I.; Munday, P.L. Shifting from right to left: The combined effect of elevated CO_2 and temperature on behavioural lateralization in a coral reef fish. *PLoS ONE* **2014**, *9*, e87969. [CrossRef] [PubMed]
36. Jutfelt, F.; de Souza, K.B.; Vuylsteke, A.; Sturve, J. Behavioural disturbances in a temperate fish exposed to sustained high-CO_2 levels. *PLoS ONE* **2013**, *8*, e65825. [CrossRef] [PubMed]
37. Sundin, J.; Jutfelt, F. Effects of elevated carbon dioxide on male and female behavioural lateralization in a temperate goby. *R. Soc. Open Sci.* **2018**, *5*, 171550. [CrossRef] [PubMed]
38. Lopes, A.; Morais, P.; Pimentel, M.; Rosa, R.; Munday, P.; Gonçalves, E.; Faria, A. Behavioural lateralization and shoaling cohesion of fish larvae altered under ocean acidification. *Mar. Biol.* **2016**, *163*, 243. [CrossRef]
39. Nilsson, G.E.; Dixson, D.L.; Domenici, P.; McCormick, M.I.; Sørensen, C.; Watson, S.-A.; Munday, P.L. Near-future carbon dioxide levels alter fish behaviour by interfering with neurotransmitter function. *Nat. Clim. Chang.* **2012**, *2*, 201. [CrossRef]
40. Rosa, R.; Pimentel, M.; Galan, J.G.; Baptista, M.; Lopes, V.M.; Couto, A.; Guerreiro, M.; Sampaio, E.; Castro, J.; Santos, C. Deficit in digestive capabilities of bamboo shark early stages under climate change. *Mar. Biol.* **2016**, *163*, 60. [CrossRef]
41. Di Santo, V.; Bennett, W.A. Effect of rapid temperature change on resting routine metabolic rates of two benthic elasmobranchs. *Fish Physiol. Biochem.* **2011**, *37*, 929–934. [CrossRef] [PubMed]
42. McLaughlin, R.H.; O'Gower, A.K. Life history and underwater studies of a heterodont shark. *Ecol. Monogr.* **1971**, *41*, 271–289. [CrossRef]
43. Rodda, K.; Seymour, R. Functional morphology of embryonic development in the port jackson shark *heterodontus portusjacksoni* (meyer). *J. Fish Biol.* **2008**, *72*, 961–984. [CrossRef]
44. Bisazza, A.; Pignatti, R.; Vallortigara, G. Laterality in detour behaviour: Interspecific variation in poeciliid fish. *Anim. Behav.* **1997**, *54*, 1273–1281. [CrossRef] [PubMed]
45. R Core Team. *R: A Language and Environment for Statistical Computing*, 3.4.3; R Foundation for Statistical Computing: Vienna, Austria, 2017.

46. Quaranta, A.; Siniscalchi, M.; Frate, A.; Iacoviello, R.; Buonavoglia, C.; Vallortigara, G. Lateralised behaviour and immune response in dogs: Relations between paw preference and interferon-γ, interleukin-10 and igg antibodies production. *Behav. Brain Res.* **2006**, *166*, 236–240. [CrossRef] [PubMed]

47. Siniscalchi, M.; Sasso, R.; Pepe, A.M.; Dimatteo, S.; Vallortigara, G.; Quaranta, A. Catecholamine plasma levels following immune stimulation with rabies vaccine in dogs selected for their paw preferences. *Neurosci. Lett.* **2010**, *476*, 142–145. [CrossRef] [PubMed]

48. Neveu, P.J. Cerebral lateralization and the immune system. In *International Review of Neurobiology*; Academic Press: Cambridge, MA, USA, 2002; Volume 52, pp. 303–323.

49. Jozet-Alves, C.; Viblanc, V.A.; Romagny, S.; Dacher, M.; Healy, S.D.; Dickel, L. Visual lateralization is task and age dependent in cuttlefish, sepia officinalis. *Anim. Behav.* **2012**, *83*, 1313–1318. [CrossRef]

50. Dharmaretnam, M.; Andrew, R.J. Age- and stimulus-specific use of right and left eyes by the domestic chick. *Anim. Behav.* **1994**, *48*, 1395–1406. [CrossRef]

51. Brown, C.; Gardner, C.; Braithwaite, V.A. Population variation in lateralized eye use in the poeciliid *brachyraphis episcopi*. *Proc. R. Soc. Lond. Ser. B Biol. Sci.* **2004**, *271*, S455–S457. [CrossRef] [PubMed]

52. Bisazza, A.; Facchin, L.; Pignatti, R.; Vallortigara, G. Lateralization of detour behaviour in poeciliid fish: The effect of species, gender and sexual motivation. *Behav. Brain Res.* **1998**, *91*, 157–164. [CrossRef]

53. Bibost, A.-L.; Kydd, E.; Brown, C. The effect of sex and early environment on the lateralization of the rainbowfish melanotaenia duboulayi. In *Behavioral Lateralization in Vertebrates: Two Sides of the Same Coin*; Csermely, D., Regolin, L., Eds.; Springer: Berlin/Heidelberg, Germany, 2013; pp. 9–24.

54. Rogers, L.J. Relevance of brain and behavioural lateralization to animal welfare. *Appl. Anim. Behav. Sci.* **2010**, *127*, 1–11. [CrossRef]

symmetry

MDPI

Article

Relationship between Motor Laterality and Aggressive Behavior in Sheepdogs

Marcello Siniscalchi [1],*, Daniele Bertino [2], Serenella d'Ingeo [1] and Angelo Quaranta [1]

[1]   Department of Veterinary Medicine, Section of Behavioral Sciences and Animal Bioethics, University of Bari "Aldo Moro", 70121 Bari, Italy; serenella.dingeo@uniba.it (S.d.); angelo.quaranta@uniba.it (A.Q.)

[2]   Med Vet, ENCI-CSEN certified professional dog trainer, 20060 Milano, Italy; danielebertino@libero.it

*   Correspondence: marcello.siniscalchi@uniba.it; Tel.: +39-080-544-3948

Received: 10 January 2019; Accepted: 11 February 2019; Published: 15 February 2019

Abstract: Sheepdogs' visuospatial abilities, their control of prey-driven behavior and their motor functions are essential characteristics for success in sheepdog trials. We investigated the influence of laterality on 15 sheepdogs' (*Canis familiaris*) spontaneous turning motor pattern around a herd and on their behavior during the first encounter with sheep in a training session. The most relevant finding of this research was that the dogs displayed significantly more aggressive behavior toward the sheep when turning in a counterclockwise direction around the herd. Considering that in counterclockwise turns the sheep were in the dogs' left visual hemifield, the high frequency of aggressions registered during counterclockwise turns suggests right-hemisphere main activation. Overall, our results revealed the existence of a relationship between motor lateralization and aggressive behavior in dogs during sheepdog training and have practical implications for sheepdog training.

Keywords: dog; behavior; laterality; vision; physiology

1. Introduction

Sheepdog trials are a worldwide sports competition involving dogs and handlers, in which the dogs' ability to manage sheep properly during different daily working activities is tested (e.g., gathering, driving, shedding, penning, and singling). Historically, sheepdogs belong to different breeds, which were selected to cooperate with humans in sheep raising, specifically in guarding and herding the animals [1]. As a consequence, the selection and training of dogs are fundamental aspects for both farm work and sheepdog trials. Among the required characteristics, the visuospatial abilities of the dogs and their control of their motor functions and prey-driven behavior are essential for the success of sheepdog activities. Predatory aggressive behavior is a part of the predation functional system, which includes different behaviors aimed at capturing and killing the prey. According to the definition of aggression, animals display aggression with the intention to do harm. Therefore, predatory aggressive behavior falls into this category, even though the animal's motivation is very different from that in other forms of aggression (e.g., social aggression) [2].

There is now clear evidence that visuospatial information is analyzed in an asymmetrical way by the dogs' brains and that it is associated with asymmetries of the dogs' motor functions [3,4]. Specifically, it has been found that agility-trained dogs displayed longer latencies to complete the weave pole obstacle (a task requiring dogs to work around pole obstacles secured in a straight line) when the owner was located in their left visual hemifield compared to when they were in the right one [3]. Given that in a dog's brain the right hemisphere neural structures are mainly fed by inputs from the left visual hemifield and vice versa (crossed fibers at the optic nerve level are more than 70% [5]), these results support the general hypothesis of right hemisphere specialization for the analysis of stimuli with high emotional valence (i.e., the owner) [6–11]. In other words, visual analysis of the owner exerted predominantly by neural pathways of the left eye (activity of the right hemisphere) was likely to increase the dogs' arousal state, distracting them during the performance of agility obstacles (resulting in longer latencies to complete the task).

A recent study has found a significant relationship between the lateralized processing of visuospatial attention and motor functions in canine species [4]. Briefly, dogs preferentially using their left paw in a motor task that required subjects to hold a puzzle feeder (namely, the "Kong test") showed a leftward bias in the total number of food items eaten from a Plexiglas board (i.e., a food detection task resembling the so-called cancellation test). Similarly, a reversed rightward bias was observed in subjects significantly preferring their right paw in the motor task; whereas, no bias was found for ambi-pawed dogs. In addition, considering the order of eating food items, the above-mentioned significant sidedness effect (left vs. right hemispace attention) was revealed only in the left-pawed group that showed a clear leftward bias (right hemisphere activity), supporting the general hypothesis of the right hemispheric superiority in spatial attention control. Apart from contributing to basic knowledge about the biology of dogs, these results could improve human abilities in canine training during different activities (animal-assisted therapy, guide dogs for the vision impaired, or sport competitions). For instance, it could be useful for a dog trainer to know the dog's visuospatial orienting bias in order to choose the best strategy to interact with it during training (e.g., to optimize the capture of his attention and/or to choose the handling side that interferes less with his orienting attention).

In light of the previous research evidence, the present study aimed at investigating the influence of laterality on sheepdog training and on their selection, evaluating the dogs' turning motor pattern around the herd and their behavior during the first encounter with sheep in a training session.

2. Materials and Methods

2.1. Subjects

Access to 15 healthy adult dogs (*Canis familiaris*: 6 males and 9 females) was obtained through an agility dog center "Mix and Breed" in Bussero, Italy. Their ages ranged from 2 to 8 years (4.6 ± 0.45; mean years $\pm$ SEM). The studied population was composed of both pure and crossbred dogs of small, medium, and large body size. Only two males and four females were neutered. None of the observed dogs had previously been trained in sheepdog activities (see Table 1 for details). In addition, the dogs were checked by a veterinarian for the presence of estrous or mouthedness (e.g., broken canines or teeth issues), which may have biased their performance under the test conditions.

Table 1. Subjects' characteristics and their behavioral score (BS) during clockwise and counterclockwise turns. The BS was computed for each subject allocating a score of 1 for each of the behaviors analyzed.

Dog	Sex	Neutered	Age (years)	Weight (Kg)	Breed	BS Clockwise	BS Counterclockwise
Ambra	female	no	6	19	Border Collie	0	6
Chobil	male	no	4	21	Border Collie	5	7
Euforia	female	no	3	23	Belgian Tervuren	14	26
Goku	male	no	4	21	Border Collie	15	4
Gulliver	male	no	4	27	Belgian Malinois	6	25
Mimi	female	no	6	24	Belgian Groenendael	11	16
Nabi	female	yes	4	22	Belgian Malinois	0	7
Smog	male	no	3	17	Mongrel	7	7
Tuli	female	no	2	22	Belgian Groenendael	3	8
Zoe	female	no	3	16	Border Collie	2	7
Geppi	female	yes	6	17	Smooth Collie	10	13
Juno	male	yes	6	16	Australian Kelpie	5	11
Ruster	male	yes	3	17	Border Collie	5	6
Vicki	female	yes	7	10	Mongrel	7	7
Vicky	female	yes	8	20	Australian Shepherd	4	6

2.2. Procedure

Each subject was led by the owner within a large enclosed area where a group of 3 sheep was placed in the center (Figure 1). As soon as the owners reached a designated position, at a distance of about 4 meters from the sheep, they unleashed the dog and left the area. The dog's interactions with the herd were recorded using two video cameras in order to have a full view of the working area. An expert sheepdog trainer was positioned close to the sheep to promptly interrupt the dog's aggression toward the sheep by using voice commands or by shaking a black plastic envelope. The recordings lasted 6 minutes, from the dog's release till the end of the working session. The dogs were initially allowed to interact freely with the sheep, turning around the herd according to their natural movement. Subsequently, they were induced by the trainer to interrupt their spontaneous rotational movement around the sheep every 20 s (if the dog kept its own rotation around the sheep for more than 20 s, the trainer stopped the movement by placing himself between the dog and the sheep). The dogs could then resume their rotation movement in the same or the reverse direction around the sheep (see Figure 1).

Figure 1. Schematic representation of the positions of the dog, the trainer, and the sheep during the session and an example of a dog making a counterclockwise turn around the herd.

During the dogs' spontaneous rotation movements around the herd, the trainer remained motionless to prevent any influence on the subjects' behavior, but he intervened just to interrupt the dogs' rotation movements (each 20 s) or to prevent aggression. The sheep were always the same during the tests, and they were familiar with sheepdog activities. The tests took place on three different days, during which 5 dogs were tested per day. The daily interactions of the dogs with the sheep were performed with 30 min intervals in order to give the sheep time to rest and recuperate.

2.3. Video Analysis

The recorded videos were scanned frame by frame by two trained experimenters. The following parameters, which included specific prey-driven behaviors, were analyzed [2]:

Turning preference: The total time (in s) spent turning clockwise and counterclockwise with respect to the sheep during the working session; and dogs' behavior: Straight approach—the dog approached the sheep along a straight trajectory, shoving—the dog pushed one sheep using its shoulder, gripping—the dog approached the sheep aggressively gripping and pulling its leg wool, approach direction—indicates the dog's approach direction toward the sheep (lateral, frontal, or backside approach), and sidedness—indicates the dogs' side (left or right) on which the sheep was positioned in the lateral approach.

The frequency of straight approaches, shoving, and gripping during the working session was computed for each dog, and the direction and sidedness of each approach was described. Therefore, the total frequency of the above-mentioned behaviors, as well as the total frequency of the different direction of the dog's approach toward the sheep (lateral, frontal, backside) were then obtained and analyzed.

2.4. Statistical Analysis

Data distribution was tested using the Shapiro–Wilk test; subsequently, the paired-samples *t*-test and the Wilcoxon signed-rank test were used to test parametric and nonparametric data, respectively. For all statistical tests, SPSS software was used, and the results were considered significant if $p < 0.05$.

2.5. Ethics Statement

The experiments were conducted in accordance with directive 2010/63/EU of the European Parliament and of the Council and were approved by the Department of Veterinary Medicine's (University of Bari) Ethics Committee, EC (Approval Number: 12/17); in addition, before the experiment began, informed consent was obtained from all the participants included in the study.

3. Results

3.1. Dogs' Turning Preference and Behavior

The relationship between lateralized turning motor pattern and sheepdog behavior is shown in Figure 2. Results revealed that gripping behavior occurred with a higher frequency during counterclockwise turns around the sheep compared to during the clockwise turns (clockwise turns: 1.06 ± 1.79; counterclockwise turns: 2.00 ± 3.07; $z = 80.00$, $p = 0.016$). In addition, when dogs ran in a circle around the flock in a counterclockwise direction, their approach to the sheep occurred significantly more along a straight trajectory (clockwise turns: 3.73 ± 0.53; counterclockwise turns: 5.33 ± 2.05; $t(14) = -2.323$, $p = 0.036$). Regarding shoving behavior, it is important to note that it occurred only on four occasions, and, in particular, three times during clockwise turns and only once during counterclockwise turns. No significant statistical differences were found in the total time spent by dogs turning clockwise and counterclockwise around the herd during sessions (clockwise: 53.54 ± 7.64 s; counterclockwise: 56.55 ± 8.47 s; $t(14) = -0.340$, $p = 0.739$). Dogs approached the sheep more frequently from their backside (87.5%) compared to their lateral (7.8%), and frontal (4.7%) sides. Regarding backside approaches, statistical analysis revealed that they occurred significantly more during counterclockwise than during clockwise turns (1.00 ± 0.37 and 2.73 ± 0.67, respectively; $z = 87.00$, $p = 0.003$). Although a slight tendency to approach sheep placed on dogs' left side was found it did not reach a statistical significance (right vs. left: Clockwise turns: $z = 18.00$, $p = 1.00$; counterclockwise turns: $z = 13.00$, $p = 0.579$; straight approaches: $z = 36.00$, $p = 0.383$; gripping behavior: $t(14) = 0.904$, $p = 0.381$).

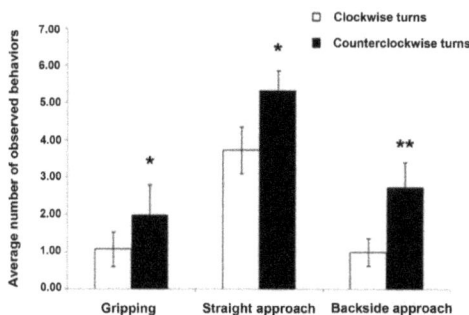

Figure 2. Dogs' behavior during clockwise and counterclockwise turns (means with SEM are shown; $* p < 0.05$; $** p < 0.01$).

3.2. Sex Ratio

The statistical analysis revealed that female dogs approached the sheep significantly more along a straight trajectory (clockwise turns: 3.22 ± 0.98; counterclockwise turns: 5.77 ± 0.52; $t(8) = -3.603$, $p = 0.007$) and from their backside (backside approach: clockwise turns: 1.11 ± 0.53; counterclockwise turns: 2.88 ± 0.97; $z = 280.00$, $p = 0.017$) during the counterclockwise turns than during clockwise turns. No differences were found for male dogs (straight approach: $t(5) = -0.139$, $p = 0.895$; backside approach: $z = 18.00$, $p = 0.096$).

No other statistically significant differences between clockwise turns and counterclockwise turns were found in male and female dogs for the other analyzed parameters: gripping (male: $z = 7.00$, $p = 0.461$; female: $z = 16.00$, $p = 0.236$) and turning preference (male: $t(5) = -0.671$, $p = 0.532$; female: $t(8) = 0.414$, $p = 0.690$).

3.3. Age

The statistical analysis showed more backside approaches by adult dogs compared to younger ones (backside approach (2–3 years: $z = 10.00$, $p = 0.059$; 4–8 years: $z = 42.50$, $p = 0.016$)). The analysis showed no other significant differences between clockwise and counterclockwise turns in the analyzed parameters according to the dogs' age: Gripping (2–3 years: $z = 4.50$, $p = 0.414$; 4–8 years: $z = 19.50$, $p = 0.344$) and straight approach (2–3 years: $z = 9.00$, $p = 0.144$; 4–8 years: $z = 28.00$, $p = 0.160$).

4. Discussion

The most relevant finding of this research was that the expression of the dogs' aggressive behavior toward the sheep is lateralized. Specifically, among the aggressive behaviors scored, the gripping and the straight approach toward the sheep occurred with a higher frequency when the dogs ran in a circle around the livestock in a counterclockwise direction. During dogs' rotational movements the sheep were viewed mainly in their monocular peripheral vision. Considering that the lateral field of each of the dogs' eyes projects mainly to the contralateral side of the brain (crossing of fibers at the optic nerve level is 70% [5]), the visual analysis of the herd by the left visual hemifield during counterclockwise turnings indicates a prevalent activation of the right hemisphere. This result fits with previous evidences about the specialization of right neural structures of dogs' brains in attending to arousal stimuli [6–13]. Previous studies employing the head-turning paradigm reported a right hemisphere main involvement in processing visually arousing stimuli, in particular the black silhouette of a cat displaying an agonistic posture (with an arched laterally displayed body and erected tail) and a snake silhouette, which is generally considered as an alarming stimulus for most mammals [7]. Moreover, a right hemisphere dominant activity was found in the dogs' response to arousing acoustic, olfactory, and visual stimuli [8–11]. Specifically, the dogs consistently turned their head with the left ear leading (right hemisphere activation) in response to thunderstorm playbacks and conspecific and human vocalizations eliciting intense emotions [10,14]. On the other hand, the dogs consistently used the right nostril (right hemisphere) to sniff cotton swabs impregnated with arousing odors (e.g., conspecific odors collected during a stressful situation, adrenaline, and veterinarian sweat) [8,9]. In addition, a right hemisphere main involvement was found in the processing of human faces expressing intense and arousing emotions (e.g., anger and fear) [11].

Recent studies on several vertebrates have reported a general specialization of the right hemisphere in the expression of intense emotions, including aggression, escape behavior, and fear [15]. In particular, our findings are consistent with the right hemisphere specialization for aggressive responses previously reported for several species, including chicks [16], horses [17], and toads [18], which showed more aggressive responses to other conspecifics when they were positioned on the animal's left side than when they were on its right.

A reasonable explanation for the reported asymmetry in dogs' aggressive behavior toward the sheep is to assume that it reflects a different activation of the two brain hemispheres during visual analysis of the target (i.e., the sheep). In addition, neurobiological studies on rats reported that the preferred direction of rotation was contralateral to the brain hemisphere with higher levels of dopamine [19], a neurotransmitter directly involved in motor control and emotional functioning.

Prey drive is a carnivore's inborn behavioral pattern to pursue and capture prey, and it is a fundamental characteristic of sheepdogs. As a matter of fact, herding behavior is modified and influenced by predatory behavior [20]. Through selective breeding, humans have been able to reduce sheepdogs' prey-driven behavior while maintaining their hunting skills. There is now evidence that the left hemisphere neural structures are better suited for the control of prey-driven behavior in dogs [21]. A similar specialization of the left hemisphere in predatory behaviors has been documented in other species, like toads [22], zebrafish [23], and black-winged stilts [24]. An interesting explanation of the prey-driven behavior control in sheepdogs during herding could be found in the inhibition exerted by the left hemisphere on aggressive behaviors, whose expression is elicited by the right hemisphere [25]. Our results support this hypothesis, since aggressive behavior occurred predominantly during the dogs' counterclockwise rotations around the sheep (right hemisphere dominant activity). Thus, it could be hypothesized that during clockwise turning, dogs controlled the herd with the prevalent use of the left hemisphere (specialized for predatory behavior), which plays a main role in sustaining the subject's attention and in risk taking, by inhibiting a fast and emotive response mediated by the right hemisphere activation [25]. This hypothesis is supported by dogs' tendency (but not statistically significant) to express "controlled" prey-driven behaviors toward the sheep during clockwise turns (left hemisphere activation). The left hemisphere activity and its functions are fundamental for the successful pursuit and capture of prey [25]. The sheep's presence led the dogs' arousal to increase, causing a right hemisphere activation, which regulates the expression of aggressive behaviors (shown in counterclockwise turns). In particular, the dogs' perception of arousing stimuli (the sheep) and the intense emotions that the sheep's presence elicited in the tested animals resulted in the dogs' aggressive response toward the herd.

On the other hand, it could be possible that the dogs' right hemisphere activation was elicited by the trainer's presence that prevented dogs from attacking the sheep. In other words, the trainer's presence could have increased the dogs' arousal levels producing a conflicting inner state caused by the dogs' ambivalent attitude to approaching the sheep (left hemisphere) and to withdrawing from them (right hemisphere) because of the trainer's presence. As a result of this conflicting situation, the dogs' stress levels (arousal) increased and the right hemisphere took control of dogs' flight or fight behavioral response.

Apart from improving the training techniques of sheepdogs, our results contribute to defining novel parameters for the assessment of animals' emotions, which could have a potential impact on their welfare. In particular, given the right–left hemisphere specializations, the evaluation of lateralized behavioral responses to an environmental stimulus, which reflects the activation of one hemisphere, could provide information about an animal's emotional state. Dominance by the right hemisphere suggests that the animal perceives the stimulus or the situation as arousing (or stressing) and potentially leads to the expression of intense emotional expressions, including aggression [16,17]. Therefore, the assessment of lateralized patterns could help to determine whether an animal experiences a certain situation or event as positive or negative and, at the same time, it could be useful to improve and ensure human safety during interactions with the dogs.

Overall our results revealed the existence of a relationship between motor lateralization and aggressive behavior in dogs during sheepdog training, supporting previous evidences about the influence of brain lateralization on visually guided motor responses in dogs. These results have direct implications for both the personnel involved in the selection of dogs to be trained for herding and for the development of new training techniques.

Author Contributions: Conceptualization, M.S., D.B., S.d., and A.Q.; methodology, M.S., D.B., S.d., and A.Q.; formal analysis, M.S., S.d., and A.Q.; investigation, M.S., D.B., S.d., and A.Q.; data curation, M.S., D.B., S.d., and A.Q.; writing—original draft preparation, M.S., S.d., and A.Q.

Funding: This research is part of the following project: "Lateralizzazione e disturbi del comportamento nel cane" supported by the University of Bari, Italy, through a financial grant to M.S.

Acknowledgments: We are grateful to Fabrizio Pertosa for technical assistance during the experiment.

Conflicts of Interest: The authors declare no conflict of interest.

References

1. The International Sheep Dog Society-Rules for Trials. Available online: https://www.isds.org.uk/trials/sheepdog-trials/rules-for-trials/ (accessed on 12 December 2018).
2. Handelman, B. *Canine Behavior: A Photo Illustrated Handbook*; Dogwise Publishing: Wenatchee, WA, USA, 2012; ISBN 0976511827.
3. Siniscalchi, M.; Bertino, D.; Quaranta, A. Laterality and performance of agility-trained dogs. *Laterality* **2014**, *19*, 219–234. [CrossRef] [PubMed]
4. Siniscalchi, M.; d'Ingeo, S.; Fornelli, S.; Quaranta, A. Relationship between visuospatial attention and paw preference in dogs. *Sci. Rep.* **2016**, *6*, 31682. [CrossRef] [PubMed]
5. Fogle, B. *The Dog's Mind*; Pelham Editions: London, UK, 1992; p. 203, ISBN 072071964X.
6. Quaranta, A.; Siniscalchi, M.; Vallortigara, G. Asymmetric tail-wagging responses by dogs to different emotive stimuli. *Curr. Biol.* **2007**, *17*, R199–R201. [CrossRef] [PubMed]
7. Siniscalchi, M.; Sasso, R.; Pepe, A.M.; Vallortigara, G.; Quaranta, A. Dogs turn left to emotional stimuli. *Behav. Brain Res.* **2010**, *208*, 516–521. [CrossRef] [PubMed]
8. Siniscalchi, M.; Sasso, R.; Pepe, A.M.; Dimatteo, S.; Vallortigara, G.; Quaranta, A. Sniffing with right nostril: Lateralization of response to odour stimuli by dogs. *Anim. Behav.* **2011**, *82*, 399–404. [CrossRef]
9. Siniscalchi, M.; d'Ingeo, S.; Quaranta, A. The dog nose "KNOWS" fear: Asymmetric nostril use during sniffing at canine and human emotional stimuli. *Behav. Brain Res.* **2016**, *304*, 34–41. [CrossRef] [PubMed]
10. Siniscalchi, M.; d'Ingeo, S.; Fornelli, S.; Quaranta, A. Lateralized behavior and cardiac activity of dogs in response to human emotional vocalizations. *Sci. Rep.* **2018**, *8*, 77. [CrossRef] [PubMed]
11. Siniscalchi, M.; d'Ingeo, S.; Quaranta, A. Orienting asymmetries and physiological reactivity in dogs' response to human emotional faces. *Learn. Behav.* **2018**, *46*, 574–585. [CrossRef] [PubMed]
12. Racca, A.; Guo, K.; Meints, K.; Mills, D.S. Reading faces: Differential lateral gaze bias in processing canine and human facial expressions in dogs and 4-year-old children. *PLoS ONE* **2012**, *7*, e36076. [CrossRef]
13. Siniscalchi, M.; Lusito, R.; Vallortigara, G.; Quaranta, A. Seeing left- or right-asymmetric tail wagging produces different emotional responses in dogs. *Curr. Biol.* **2013**, *23*, 2279–2282. [CrossRef] [PubMed]
14. Siniscalchi, M.; Quaranta, A.; Rogers, L.J. Hemispheric specialization in dogs for processing different acoustic stimuli. *PLoS ONE* **2008**, *3*, e3349. [CrossRef] [PubMed]
15. Rogers, L.J.; Andrew, R. *Comparative Vertebrate Lateralization*; Cambridge University Press: Cambridge, UK, 2002; ISBN 0-521-78161-2.
16. Vallortigara, G.; Cozzutti, C.; Tommasi, L.; Rogers, L.J. How birds use their eyes: Opposite left-right specialization for the lateral and frontal visual hemifield in the domestic chick. *Curr. Biol.* **2001**, *11*, 29–33. [CrossRef]
17. Austin, N.P.; Rogers, L.J. Limb preferences and lateralization of aggression, reactivity and vigilance in feral horses, Equus caballus. *Anim. Behav.* **2012**, *83*, 239–247. [CrossRef]
18. Robins, A.; Rogers, L.J. Lateralized prey-catching responses in the cane toad, Bufo marinus: Analysis of complex visual stimuli. *Anim. Behav.* **2004**, *68*, 767–775. [CrossRef]
19. Carlson, J.N.; Glick, S.D.; Hinds, P.A.; Baird, J.L. Food deprivation alters dopamine utilization in the rat prefrontal cortex and asymmetrically alters amphetamine-induced rotational behavior. *Brain Res.* **1988**, *454*, 373–377. [CrossRef]
20. Renna, C.H. *Herding Dogs: Selection and Training the Working Farm Dog*; Kennel Club Books (KCB): Allenhurst, NJ, USA, 2008; ISBN 978-1-59378-737-0.
21. Siniscalchi, M.; Pergola, G.; Quaranta, A. Detour behaviour in attack-trained dogs: Left-turners perform better than right-turners. *Laterality* **2013**, *18*, 282–293. [CrossRef] [PubMed]
22. Lippolis, G.; Bisazza, A.; Rogers, L.J.; Vallortigara, G. Lateralization of predator avoidance responses in three species of toads. *Laterality* **2002**, *7*, 163–183. [CrossRef] [PubMed]

23. Miklósi, Á.; Andrew, R.J.; Savage, H. Behavioural lateralisation of the tetrapod type in the zebrafish (Brachydanio rerio). *Physiol. Behav.* **1998**, *63*, 127–135. [CrossRef]

24. Ventolini, N.; Ferrero, E.A.; Sponza, S.; Della Chiesa, A.; Zucca, P.; Vallortigara, G. Laterality in the wild: Preferential hemifield use d uring predatory and sexual behaviour in the black-winged stilt. *Anim. Behav.* **2005**, *69*, 1077–1084. [CrossRef]

25. Rogers, L.J.; Vallortigara, G.; Andrew, R.J. *Divided Brains: The Biology and Behaviour of Brain Asymmetries*; Cambridge University Press: Cambridge, UK, 2013; ISBN 0521540119.

symmetry

MDPI

Article

Laterality as a Predictor of Coping Strategies in Dogs Entering a Rescue Shelter

Shanis Barnard *, Deborah L. Wells and Peter G. Hepper

Animal Behaviour Centre, School of Psychology, Queen's University Belfast, Belfast BT7 1NN, UK;
d.wells@qub.ac.uk (D.L.W.); p.hepper@qub.ac.uk (P.G.H.)
* Correspondence: shanis.barnard@gmail.com

Received: 14 September 2018; Accepted: 19 October 2018; Published: 23 October 2018

Abstract: It has been reported that during the first few days following entry to a kennel environment, shelter dogs may suffer poor welfare. Previous work suggests that motor bias (the preferred use of one limb over the other) can potentially be used as an indicator of emotional reactivity and welfare risk. In this study, we investigate whether paw preference could be used as a predictive indicator of stress coping (measured using cortisol levels and behavioural observation) in a sample of 41 dogs entering a rescue shelter. Cortisol levels and behavioural observations were collected for one week after admission. We scored the dogs' paw preference during a food-retrieval task. Our results showed that increasing left-pawedness was associated with a higher expression of stress-related behaviours such as frequent change of state, vocalisations and lower body posture. These results are in keeping with previous findings showing that left-limb biased animals are more vulnerable to stress. Paw preference testing may be a useful tool for detecting different coping strategies in dogs entering a kennel environment and identifying target individuals at risk of reduced welfare.

Keywords: dog; laterality; paw preference; shelter; welfare

---

1. Introduction

Entering a rescue shelter can be a very stressful experience for dogs; they are separated from any social attachment figures, they are exposed to a novel environment (i.e., unfamiliar noise, smells, disruption of familiar routine) and to daily interactions with unfamiliar people and conspecifics [1–3]. Previous studies show how this transition can generate a state of fear, anxiety, and frustration [4,5]. Social and spatial restriction following confinement can be a cause of both acute and chronic stress [6–8]. Physiological studies have confirmed that kennelling is perceived by dogs as a psychogenic stressor, with animals displaying peaks in cortisol levels in the first few days after arrival [3,5,9]. Individual dogs, however, may have different coping abilities or stress resilience. Previous research has highlighted two main coping styles in individuals that are environmentally challenged: proactive and reactive [10]. Proactive coping styles may be more typical of 'bold' personality types and are defined by active attempts to counter the stressful stimuli and by low Hypothalamic-Pituitary-Adrenal (HPA)-axis reactivity. Reactive, or passive, coping may be more typical of 'shy' personality types, and involves higher activation of the HPA-axis system, immobility, and low levels of aggression [11,12]. Hiby and colleagues [13] found that dogs showing or not showing physiological adaptation to kennelling had different behavioural styles, that is, dogs with lower HPA-axis activation spent more of their time walking or trotting compared to dogs with high cortisol levels. Effective interventions to minimise signs of poor welfare should be based on the evaluation of dogs' individual ability to cope and adapt to confinement in kennels. Behavioural indicators offer excellent validated measures of welfare as they represent the output of a range of sensory and cognitive experiences and decision-making processes, reflecting the expression of the animal's underlying emotional and physiological state [14].

Unfortunately, behavioural analysis, live or from a video, has several limitations and disadvantages, including being very time consuming and requiring a large amount of labour, thereby limiting the amount of information that can be collected.

Recently, when evaluating the impact of the environment on confined animals, animal welfare scientists have focused their attention on the expression of positive emotions and on the link between emotional stress and cognitive processes [15–18]. The identification of reliable and practical cognitive indicators of emotional distress would allow us to target interventions aimed at improving dogs' quality of life.

Laterality has been used as a measure of emotionality, stress reaction, and temperament in different species [18–23]. Emotional informations are processed differently by the brain hemispheres according to their valence. It has been suggested that withdrawal-related emotions are processed and controlled primarily by the right hemisphere, while positive, approach-related emotions are controlled mainly by the left hemisphere [24,25]. However, the overall expression of intense emotions, independent of their valence, has been associated, by other scholars, to a right hemispheric dominance [26,27]. Behavioural laterality may reflect this divergent hemispheric processing. Higher emotional indices have been associated with right hemisphere activation in horses [28]. Lateralisation has been reported to be linked with the intensity of behavioural reactions in novel situations [23] and to boldness in exploring novel objects and environments [29,30]. Consistent individual behavioural differences in, for example, limb preference for simple reaching, may be linked to a dominant control of the contralateral brain hemisphere. This allows the use of behaviour as an indicator of brain laterality [20]. Right-handed marmosets (left hemispheric dominance), for example, were found to be more bold, readily approaching and exploring for longer a novel object in an unfamiliar environment compared to left-handed marmosets (right hemispheric dominance) [29]. A study by Batt and colleagues [31] found that a greater strength and directionality of laterality were linked with more confident and relaxed behaviour in dogs that were exposed to novel stimuli and unfamiliar environments. Thus, laterality appears to be a potential novel indicator of coping abilities and vulnerability to stress for domestic dogs entering a kennel environment.

Laterality in dogs has been largely studied in the form of paw preference [32–34]. In this study, we assessed, for the first time, the relationship between canine laterality, as determined by the commonly used Kong ball test [23,33,35,36], and behavioural and physiological measures of stress in dogs admitted to a rescue shelter. The aim was to determine whether laterality could be used as a potential predictor of welfare risk in kennelled dogs.

2. Methods

2.1. Animals

Data collection was performed at the Dogs Trust Rehoming Centre in Ballymena, Co. Antrim, UK, over a period of 9 months. All dogs entering the shelter during this period were enrolled in the study; exceptions were made if the dog was pregnant, seriously ill or injured, impossible to handle or walk on the lead due to excessive fear or aggression. A total of 41 dogs were assessed, 22 males (54.5% neutered) and 19 females (58.0% spayed), including a number of different purebreds and crossbreeds. The minimum age for a subject to be enrolled was 12 months; the oldest dog in our sample was 9 years (median = 3 years; mean $\pm$ SD 3.7 $\pm$ 2.5 years) (Table 1).

Table 1. Demographics of dogs included in the study.

Dog ID	Name	Breed	Sex	Age (Years)	Castration
1	Roxy	TerrierX	F	5	yes
2	Ginger	TerrierX	F	5	yes
3	Albert	LhasaApso	M	4.5	yes
4	Jojoe	BeardCollie	M	5	yes
5	Dandy	Poodle	M	6	yes
6	Curly	Poodle	M	6	yes
7	Lucy	Lab	F	1.5	yes
8	Bailey	KingCharles	M	5	no
9	Ermet	JackRuss	M	1.4	yes
10	Leo	JackRussX	M	1	no
11	Tiny	MinShetland	F	8	no
12	Orsha	CockerSp	F	2	yes
13	Rex	Springer	M	1	no
14	Socks	RughCollie	M	1.5	yes
15	Roy	Collie	M	3	yes
16	Dappy	Pugalier	M	1.5	no
17	Prince	Lab	M	4	no
18	Darcy	Husky	F	3	yes
19	Svaras	Lab	M	2.5	no
20	Roxy	LabX	F	3	yes
21	Milly	CarinTerr	F	9	yes
22	Willy	CarinTerr	M	9	yes
23	Biddy	FoxTerr	F	5	yes
24	Maggie	Jug	F	1.5	no
25	Missy	IrishTerr	F	9	yes
26	Maya	LabPitX	F	4	no
27	Harley	Dobie	M	1	no
28	Dax	Dasch	M	2	no
29	Miles	Dasch	M	2.5	yes
30	Max	IrishWater	M	1.5	no
31	Charlie	Collie	M	1	yes
32	Tess	PatterdaleX	F	1	yes
33	Zoe	JackRuss	F	2	no
34	Tia	Shihtzu	F	8	no
35	Gizmo	Maltese	M	6	no
36	Lily	Shorky	F	2.5	no
37	Benson	CollieX	M	1.5	yes
38	Honey	Yorkie	F	7	no
39	Coco	TerrierX	F	5	no
40	Enzo	JackRuss	M	2.5	yes
41	Aria	JackRuss	F	2	yes

A portion of dogs (14.6%, unknown history), were rescued from the municipal pound or other shelters and the remainder were surrendered to the Dogs Trust by their owners for a variety of reasons. Main reasons included family health issues (e.g., allergies or illness of a family member), work commitments (i.e., not being able to take care of the dog anymore), owner's death, or moving home as most common (Table 2).

Table 2. Reasons provided by the owners when surrendering their dogs to the rescue centre.

Reason for Surrender	Cases	%
Family health	9	22.0
Work commitments	7	17.1
Pound/other shelters	6	14.6
Owner's death	5	12.2
Moving home	4	9.8
Behavioural problems	2	4.9
Handover/rescued	2	4.9
Problems with other household dogs	2	4.9
Too many dogs in household	2	4.9
From breeding stock	1	2.4
New born baby	1	2.4
Total	41	100

2.2. Procedure

On the day of admission to the shelter (Day 0), dogs were enrolled and general information collected from shelter records (sex, age, provenience, date of entrance, etc.). Data collection started on the following morning (Day 1) for one week, with sampling occurring on days 2, 3, 5 and 7. On each of the sampling days, both urine samples and behavioural recordings were collected following the protocols described below. Paw preference was assessed once on Day 3, see later (Section 2.2.3). The day of paw preference assessment was decided during pilot testing: it was noticed that dogs were not interested in the toy during their first two days in the kennels, whereas on day three most dogs would actively interact with, and retrieve food from, the ball.

2.2.1. Urine Cortisol/Creatinine Ratio

Between 08:00 h and 10:00 h of sampling days (1, 2, 3, 5, and 7), dogs were leashed and walked outdoors; a mid-stream sample of naturally voided urine was collected using urine sampling kits (Rocket® URIPET™, Washington, WA, USA). Urine tubes were labelled and stored at −30 °C for up to 3 months. Samples were shipped in batches to IDEXX laboratories (West Yorkshire, UK) and tested for cortisol/creatinine (C/C) ratio. Cortisol was extracted with chemiluminescent competitive immunoassay using the Siemens Immulite 2000, whereas creatinine with Jaffe (alkaline picrate) reaction using the Beckman AU 5800 analyser.

2.2.2. Behavioural Observations

Behavioural observations started after urine sampling (Days 1, 2, 3, 5, and 7). A digital camera on a tripod was positioned in front of the kennel and recorded the behaviour of each dog over three sessions of 35 min (30 central minutes extracted for analysis). Members of staff prioritised cleaning the kennels and feeding the dogs that were going to be recorded on that day. This allowed no interruptions during recording. Routine activities, however, continued as normal in the adjacent kennels. Observational sessions were distributed as follows: Session 1 (OB1) between 09:00 h and 10:00 h, during morning activities. Session 2 (OB2) between 11:00 h and 12:00 h during staff tea break; Session 3 (OB3) between 13:00 h and 15:00 during visiting hours (the Centre opened to the public between 12:00 and 16:00 h). The sessions were planned this way to allow for an overview of the range of behaviours that dogs performed across a typical day.

2.2.3. Paw Preference Assessment

Dogs' paw preferences were assessed on Day 3 using the 'Kong ball test', one of the most commonly used measures of canine motor bias [33,35,36]. Following previously published protocols, each dog was allowed to sniff a Kong Ball™ (KONG Company, Golden, CO, USA; a hollow conical-shaped toy) filled with dog food. Then the Kong was placed inside the kennel on the floor centrally in front of the animal. The paw used (left, right) by the dog to hold/stabilise the toy was recorded using a smartphone app purposely developed for this study. The test continued until 100 paw uses (left or right) had been made. The use of both paws was recorded but not counted towards this total. The use of the app had several advantages, including not having to interrupt eye contact with the subject to write down the scores, recording would automatically stop when the 100 paw-use target was reached, main statistics for the dog appeared immediately on the screen, and raw data were readily available for download in Microsoft Excel format. A single paw use was recorded regardless of how long the paw stayed on the ball. The animal was required to remove its paw completely from the ball for paw use to be scored as a separate response. Dogs were tested for approximately 30 min (the average length of time taken to collect 100 data points).

2.3. Analysis

All analyses were carried out using IBM SPSS Statistics 21.0 (IBM, Armonk, NY, USA).

2.3.1. Paw Preference Assessment

Individual laterality scores were calculated using a binomial test and converted to a z-score using the formula $z = (L - 0.5N)/\sqrt{(0.25N)}$, L being the number of left paw uses and N the total of left and right paw uses. A z-score ≥ 1.96 indicates a left paw bias, a z-score ≤ -1.96 indicates a right paw bias; a value between these two scores indicates no lateral bias (ambilateral) [33,36]. A chi-squared test was used to calculate departures from random distribution of left-, right- and ambilateral paw preference groups. Paw-preferent (either to the left or right) vs. ambilateral, and right- vs. left-paw preferent animals were compared using binomial tests to assess any significant group difference. Chi-squared tests were also used to assess whether the pawedness classification was associated with the dogs' sex (male, female) or castration status (neutered, intact).

A directional laterality index (LI) was calculated to quantify each dog's paw preference on a continuum from strongly left-paw preferent (+1) to strongly right-paw-preferent (−1). The LI score was calculated as $(L - R)/(L + R)$, where R represents the number of right paws and L the number of left paws used [30]. A score of 0 indicates no bias, a score of ±1 indicates that the subject used the same paw throughout the trial. In addition to the directional bias of lateral behaviour (i.e., left or right bias), the strength of laterality has also been used as a proxy measure of hemispheric brain activity. Strongly lateralised animals show a greater activity of one hemisphere (irrespective of the side), while weakly lateralised animals do not show a significant dominance of one hemisphere over the other (i.e., ambilateral) [25]. The absolute value of LI (LI_ABS) gives a measure of the strength of laterality, irrespective of the direction of paw use. A Shapiro-Wilk normality test was used to assess the distribution of LI and LI_ABS values to identify any population bias. Any effect of sex or castration status on the direction and strength of laterality was calculated using a Mann-Whitney-U test for independent samples.

2.3.2. Behavioural Analysis

A total of 307.5 h of footage was analysed. Videos were scored using the behavioural recording software The Observer XT13 (Noldus, The Netherlands). Due to time and resource constraints we used an instantaneous sampling method. However, given that some behaviours (e.g., barking) might have been lost or underestimated using this approach, we also scored a subset of videos using continuous sampling (Table 3). Instantaneous sampling was used every 60 s: the observer recorded the behaviour expressed by the focal animal at a given instant and scoring was performed for all the video footage (i.e., 90 data points per observational day, 450 data points for the 5 days). Continuous sampling was performed during the first 15 min of recoding of session OB1 on each of the 5 sampling days.

Table 3. Behavioural variables recorded during the behavioural observations either using instantaneous (i) or continuous (c) sampling techniques. During the latter, behaviours could be scored as either duration (d) or frequency (f).

Label	Behaviour Description	Sampling Method
Inactive behaviours	Stand: the dog is still, standing on fours	(i)
	Sit: the dog is sitting on hind legs	(i)
	Lie down: the dog is in a recumbent position with the head up (vigilant)	(i)
	Rest/sleep: the dog is lying on the floor or curled up in the bed with the head also on the ground, likely eyes are closed although not always visible.	(i)
Posture	High: the tail and/or the head are held high and the ears are forward and/or the animal is standing in an elevated posture compared to the neutral breed posture	(i)
	Neutral: breed specific posture in neutral conditions (mouth, ears and tail are relaxed)	(i)
	Low: lowered positions of the tail (or tail curled forward between the hind legs) and/or bent legs and/or backward positioning of the ears compared to neutral conditions	(i)
Moving	Walking: the dog is moving step by step with a normal pace (compared to its breed and size)	(i)
	Trotting: the dog is moving with a faster gait compared to walking	(i)
Change of state	Scored each time the dog changed from one inactive behaviour to another (e.g., sit to lie down) or from an inactive behaviour to an active one (e.g., stand to walk)	(c) (f)
Vocalisations	Barking: short staccato vocalization	(c) (f)
	Howling/whining: the dog is howling or whining	(c) (f)
	Growling: the dog is growling	(c) (f)
Socialising (with pen mate)	Positive: affiliative behaviour, playing behaviour and/or greeting behaviour	(c) (d)
	Negative: threatening behaviour and/or aggression	(c) (d)
Repetitive behaviour	Presence of stereotypical behaviour (e.g., pacing, circling, repetitive jumping on the fence, tail chasing or any other behaviour repeated in the same way for several times without any seeming function) throughout the test phase	(c) (d)
Stress-related behaviour	Behaviours that are potentially related to stressful conditions, in particular licking lips, panting, drinking, grooming/scratching, body shaking, digging at walls, doors or floor, paw lifting, yawning and startling	(c) (f)
Play	Individual playing behaviour (e.g., grabbing or holding an object in the mouth and head-shaking, play bow)	(c) (d)
Other	Scratching: Dog is scratching the floor or the walls	(c) (d)
	Rear wall: Dog rears up against the window or side walls, including bouncing in an excited manner	(c) (d)
	Drinking: Dog is drinking water from the bowl	(i)

2.3.3. Predictive Factors Affecting Behavioural Variations

To investigate if any variation in behaviour could be associated with the observation time, days from shelter entrance, sex, castration status, C/C ratio or laterality (either direction or strength), we performed Linear Mixed Model (LMM) analysis with each behaviour as the dependent variable and sex, castration status, observation times, day, C/C, LI and LI_ABS as predictive factors and dog identity as random.

2.3.4. Laterality as a Predictive Factor of C/C Level Variation

To investigate if any variation in cortisol levels throughout the observation time could be predicted by laterality (either direction or strength), we performed LMM with C/C ratio as the dependent variable and day of observation as the repeated measure, LI and LI_ABS as predictive variables, and dog identity as a random factor.

3. Results

3.1. Paw Preference Assessment

When calculating lateralisation at the individual level, 8 (19.5%) dogs mainly used their left paw to hold the Kong, 18 (43.9%) mainly used their right paw and 14 (34.1%) used both paws equally. Binomial tests showed no significant difference between the distribution of lateralised and ambilateral dogs ($p = 0.81$) or between the number of left and right pawed dogs ($p = 0.80$). Paw preference was not successfully recorded for one dog (2.5%).

The distribution of the three pawedness classification groups did not differ significantly from that expected by chance, that is, no population level effect ($\chi^2_{2,40} = 3.80$, $p = 0.15$).

No significant association emerged between dogs' paw preference classification and sex or castration status ($\chi^2_{2,40} = 1.48$, $p = 0.48$; $\chi^2_{2,40} = 3.98$, $p = 0.14$ respectively).

No population bias was recorded when exploring either the direction of laterality (using LI scores) or the absolute strength of laterality ($W = 0.95$, $p = 0.59$; $W = 0.96$; $p = 0.18$ respectively).

Direction and strength of laterality were not significantly affected by the sex of the dogs or their castration status (Sex: $Z_{LI} = -0.03$, $p = 0.98$; $Z_{|LI|} = -1.53$, $p = 0.13$; Neutering: $Z_{LI} = -0.52$, $p = 0.60$; $Z_{|LI|} = -1.19$, $p = 0.23$).

3.2. Behavioural Data Management

Given that the different sampling methods (instantaneous and continuous) were performed for different lengths of time, analyses were kept separate using two different datasets. From an initial inspection, a few behaviours were performed only rarely or by a small number of animals. Only behaviours with a median >0 were included in the analysis. Behaviours excluded were: play, socialising, repetitive behaviours, stress-related behaviours, and all behaviours under 'other'. Walking and trotting were analysed as one variable termed 'moving', while barking, howling/whining, and growling were analysed as one variable termed 'vocalisations'.

3.3. Predictive Factors Affecting Behavioural Variations

The dogs' behaviour did not vary greatly over their first week in the rescue shelter (Table 4), with the exception of 'posture', with dogs showing an increase in higher posture over time (D1 = 2.7 ± 5.4; D3 = 4.2 ± 6.9; D7 = 5.6 ± 6.9). The time of observation had a significant effect on the expression of behaviours including standing, resting/sleeping, and maintaining a low posture (Table 4). Descriptive analysis showed that dogs spent less time standing during the second bout of observation t (OB2) (i.e., staff break, 8.1 ± 6.9) compared to OB1 (i.e., cleaning, 12.3 ± 8.1) and OB3 (i.e., visiting hours, 12.0 ± 7.8). On the contrary, dogs spent more time resting/sleeping during the second observation bout (OB2) (7.4 ± 8.4) compared to OB1 (4.4 ± 7.2) and OB3 (4.2 ± 6.1). Finally, dogs were likely to show less low posture during the second observation bout (OB2) (4.1 ± 5.6) compared to OB1 (6.4 ± 7.4) and OB3 (6.4 ± 7.4). These outcomes are rather intuitive as dogs were undisturbed, thus mainly resting during OB2 and scored as having neutral posture.

Table 4. Linear Mixed Models output. Dependent variables (i.e., behaviours) on the row and predictive variables on the columns. Cells show F-values and significance levels.

Variable	df	Day	OB	N/S	Sex	C/C	LI	ABS_LI
Stand	419	0.62	16.9 ***	17.5 ***	4.5 *	15.6 ***	0.34	35.3 ***
Rest/Sleep	419	0.98	7.6 **	1.0	0.12	0.57	12.5 ***	27.0 ***
Lie down	419	1.4	1.8	34.3 ***	4.4	28.9 ***	8.0 **	6.5 *
Sit	419	1.5	0.7	0.05	0.002	8.7 **	53.7 ***	0.6
Moving	419	1.3	1.4	7.6 **	0.6	10.3 **	11.7 **	1.7
Low posture	419	2.05	5.2 **	6.5 *	0.005	1.6	7.2 **	24.1 ***
High posture	419	2.6 *	2.5	3.9	0.03	25.6 ***	3.01	7.8 **
Change posture	133	0.5	NA	0.7	0.04	1.6	10.7 **	0.4
Vocalisations	133	0.6	NA	1.3	3.8	0.9	14.4 ***	19.4 ***

df = degrees of freedom; OB = observation time; N/S = neuter/spay; C/C = cortisol/creatinine ratio; LI = laterality index; ABS_LI = strength of laterality; * $p < 0.05$; ** $p < 0.001$; *** $p < 0.0001$.

Sex and castration status had an effect on the expression of a few behaviours (Table 4); descriptive analysis highlighted that male dogs spent slightly more time standing (11.1 ± 7.4) than female dogs (10.3 ± 8.3). Entire dogs spent more time standing (11.4 ± 8.2), less time moving (1.5 ± 1.8), and less time lying (3.4 ± 4.8) than neutered/spayed dogs (10.2 ± 7.6; 2.6 ± 3.4; 5.2 ± 6.2 respectively).

Variations in cortisol levels were associated with variation in behaviour expression (Table 4). Dogs with increasingly higher cortisol levels were more likely to spend time inactive (sitting and lying down) but less time moving, standing, and showing a high posture (Table 4).

LMM analysis highlighted a significant effect of both LI and ABS_LI on a number of behaviours (Table 4). Increasing left pawedness was associated with more frequent change of state, more time spent vocalising (Figure 1a), sitting (Figure 1b), and showing a lower posture (Figure 1c). By contrast, dogs with increasingly strong right-paw bias were likely to spend more time resting/sleeping, lying down (Figure 1d) and moving.

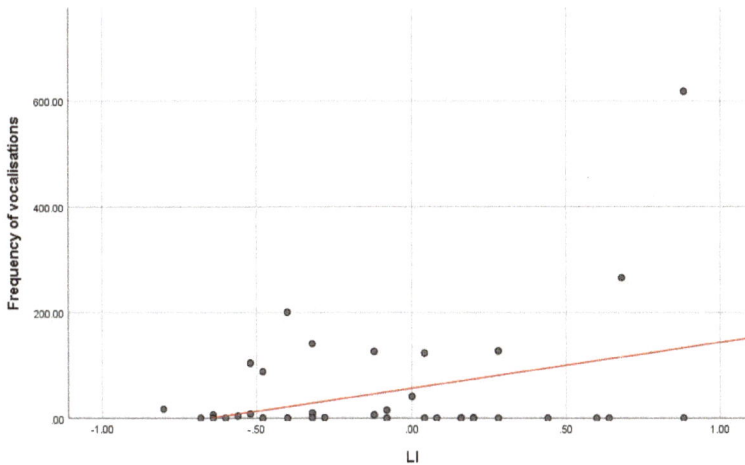

(a)

Figure 1. *Cont.*

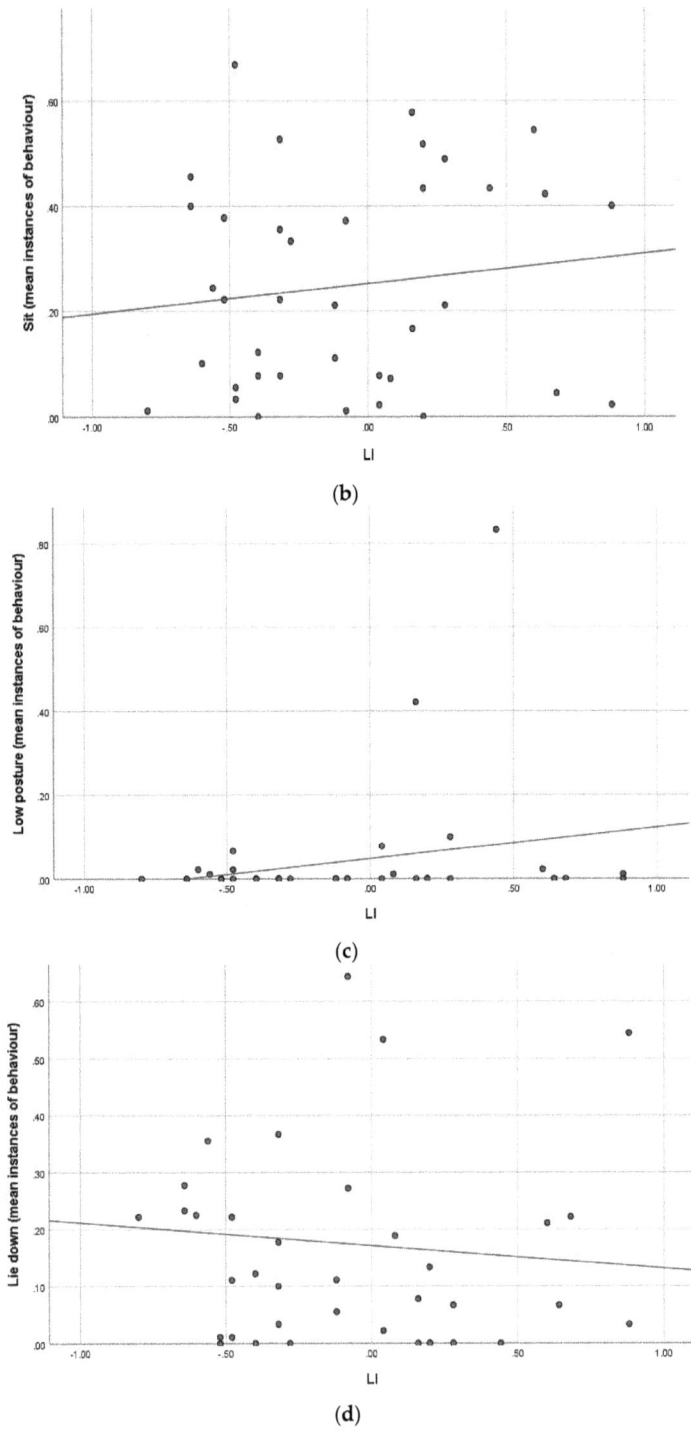

Figure 1. Regression between laterality index (LI) and vocalising (**a**), sitting (**b**), showing a low posture (**c**), and lying down behaviour (**d**). LI = −1 strong right-pawed bias; LI = 1 strong left-pawed bias.

Having a weaker lateralisation was predictive of spending less time lying down, resting/sleeping (Figure 2a) or vocalising (Figure 2b), and more time standing (Figure 2c) and showing a low or high posture.

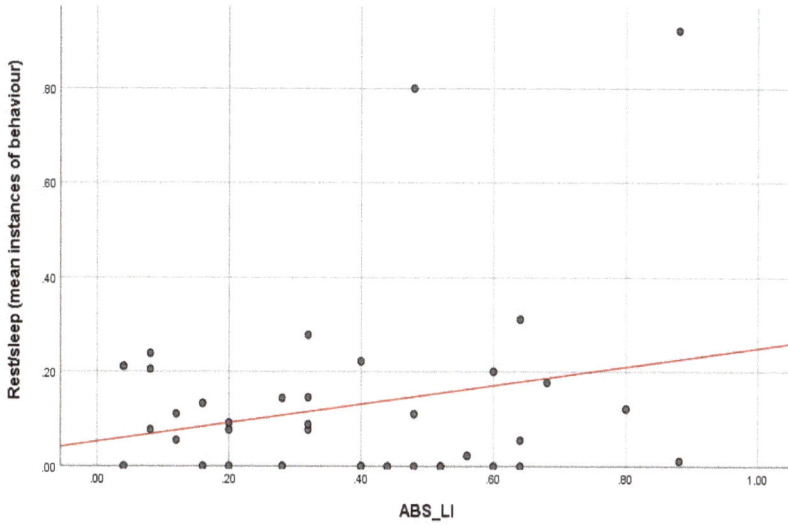

(a)

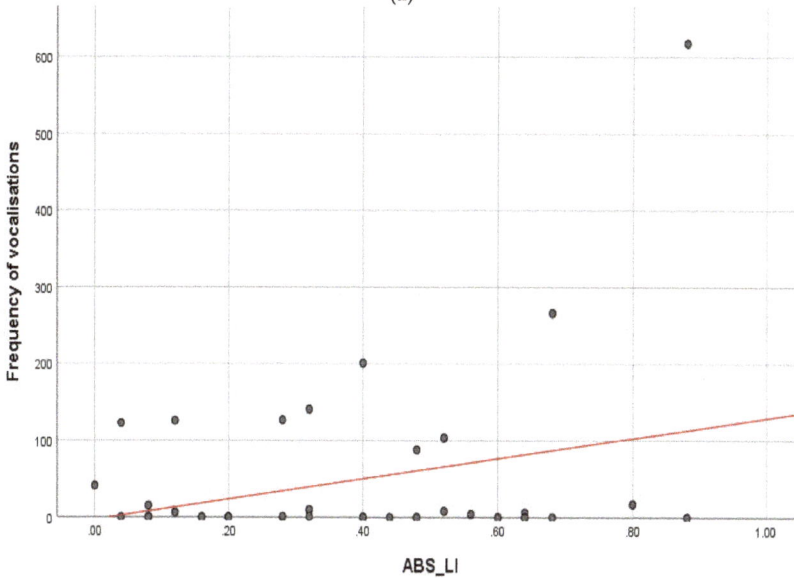

(b)

Figure 2. *Cont.*

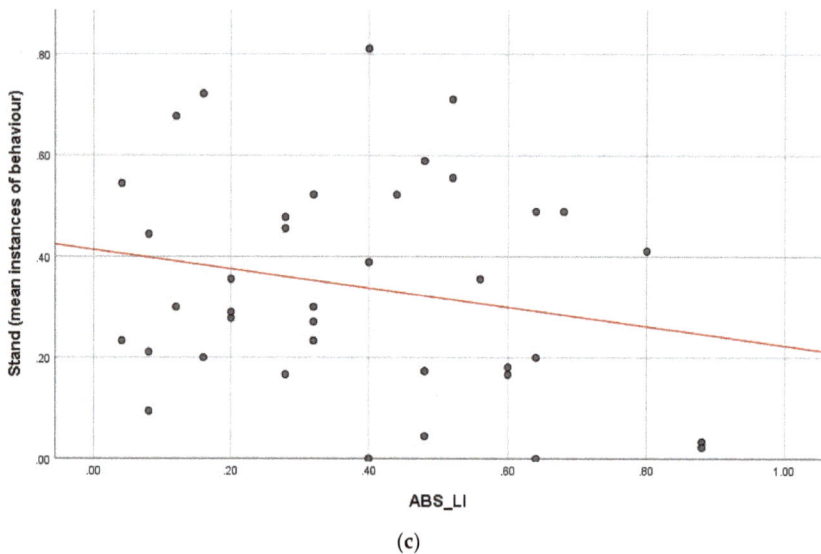

(c)

Figure 2. Association between ABS_LI and rest/sleep (**a**), vocalization (**b**) and standing (**c**). ABS_LI = 0 weak laterality; ABS_LI = 1 strong laterality.

3.4. Laterality as a Predictive Factor of C/C Level Variation

LMM showed no significant effect of laterality (direction or strength) on the levels of C/C ratio in dogs ($p > 0.05$).

4. Discussion

This study provides the first evidence for the potential use of paw preference as an indicator of coping strategy in dogs entering a kennel environment.

At the individual level, results from the Kong tests revealed a significant paw preference bias (either left or right, 63.4%), but no significant population bias emerged. This supports some previous research in this area [23,35,37]. However, results seem to depend greatly on the population of animals under study, with other authors having reported population biases in dogs [36,38]. Neither the direction nor strength of the dogs' paw preferences differed significantly according to canine sex or castration status. Again, the literature is divided on this matter and results appear to be population-dependant [23,32,33,35–37].

Our results showed a relationship between the dogs' paw preferences and certain behaviours. Right-pawedness was associated with a frequent expression of both active and inactive behaviours such as moving or resting/sleeping and lying down. Increasing left-pawedness, by contrast, was associated with a higher expression of stress-related behaviours, including frequent change of state, vocalisations, sitting, and low posture. These results tie in nicely with previous findings of a right hemisphere dominance (left motor bias) in animals being more vulnerable to stress [20]. For example, when stressed by being caged and placed in an unfamiliar environment, cats with higher cortisol levels also had a higher activation of the right brain hemisphere [39].

In a review paper, Rogers [20] suggested that the left hemisphere has a dominant control of proactive/calm behaviours (e.g., approach, exploration) and the right hemisphere of reactive behaviours (e.g., fear, aggression). In marmosets, for example, studies showed that left-handed animals (right hemispheric dominance) were more fearful of alarm calls [40], more likely to show higher cortisol levels [41], and were less reactive to the effects of social facilitation on capturing prey [42] than right-handed marmosets (left hemispheric dominance). Thus, it could be hypothesised that animals with a left-limb bias are less likely to exploit new resources and are more likely to express negative emotional functioning compared to individuals with a right-limb bias [32]. Recent studies on marmosets [43] and dogs [44] confirmed this link by observing that a stronger left-limb motor lateralisation was associated with a more negative or 'pessimistic' cognitive bias.

Dogs may show different behavioural styles according to their physiological adaptation. In this study, we found that lower cortisol levels were associated with more frequent observations of moving, standing, and high posture. It could be argued that dogs showing these behaviours were better at coping with the kennel environment, showing a more confident posture and being more active. Hiby and colleagues [13] found that on days when dogs were more active, cortisol levels were lower compared to days when dogs were spending more time sitting or lying down. High activity levels do not necessarily indicate lower stress; there is evidence of increased activity following social and spatial restrictions [7]. It could be the case that staying active helps dogs to cope better with confinement, hence the lower cortisol levels [13]. Interestingly, our results showed that a stronger right-paw bias was linked to higher activity. This lends support for the previously suggested association between left-hemispheric bias and exploratory/proactive coping styles [20].

A high cortisol profile was associated with increased observations of sitting and lying down. Once more these results are in line with findings by Hiby et al. [13]. Both behaviours indicate a more vigilant posture (compared to e.g., resting/sleeping) and a more reactive posture. Following the proactive/reactive hypothesis, dogs showing more sitting and lying behaviour are expected to be more strongly left-pawed. This was only partially true, as we found left-biased dogs did indeed spend more of their time sitting (Figure 1), but less time lying down and resting/sleeping. The link between stress, physiology, and behaviour is complex and still debated [3,5,45]. Multiple factors may play a role in modulating the results, including individual variability and past experiences [7,13,45]. Here, we added a new indicator of emotional state (as assessed by motor bias) that could help triangulate our results. If, based on previous work, we consider right-paw biased animals to be less susceptible to environmental stress, we could suggest that animals that either are very active or, on the contrary, very quiet, might cope better overall with novel environmental challenges. Left-pawed dogs, typically reacting to kennelling by showing anxiety and stress-related behaviours such as vocalising, low posture, sitting, and frequently changing state might, on the contrary, find adapting to confinement more challenging.

Strength of laterality (rather than the direction of an animal's motor bias) has been previously associated with increased levels of stress behaviour. In dogs, for example, weaker motor laterality has been linked to higher reactivity and fear of thunderstorms and firework sounds when compared to stronger lateralisation [36]. Our results show that weaker lateralisation was associated with a higher occurrence of standing behaviour and maintenance of a low and high posture. Both high and low posture may be a sign of strong emotional arousal, with different valence (e.g., aggression/extraversion versus fear respectively). Previous work has found that weak limb preference was in fact associated with more fearful, as well as excitable, reactions [31,36]. Having a strongly lateralised brain, by contrast, seems to be advantageous for enhanced cognitive abilities and higher survival fitness [46]. In our study, resting/sleeping and lying down was linked to a higher degree of lateralisation. As mentioned for the right-paw bias, it could be that spending most of the time sleeping helps an individual to cope better with the stress of entering a kennel environment. It is worth mentioning, however, that spending large amounts of time sleeping has also been reported in sheltered dogs as an indicator of learned helplessness and depression-like state [4] and would be more typically associated with a reactive coping style [10].

We found no association between cortisol levels and laterality. The right hemisphere has been found to have a dominant control of endocrine function, especially of the HPA axis [20]. Being associated with more fearful behaviour, left-limb bias should, in theory, be associated with higher corticoid responses than right-limb bias, as found in rhesus macaques and rats [47,48]. In dogs, Siniscalchi and collaborators [49] found a chronic elevation of hair cortisol in those individuals showing a higher reactivity to acoustic stimuli with different emotional valence. However, Batt and colleagues [31] found no correlation between behaviour, laterality, and salivary cortisol in dogs. This is something that should perhaps be explored further.

Overall, it appears that a left-motor bias may be linked to a more negative affective state, a more reactive coping style, and a more challenging adaptation to novel environments. Assessing paw preference may become a useful tool to detect different coping strategies in dogs entering a kennel and reduce stress in target individuals at higher welfare risk. Further work is needed to explore this in greater depth.

Author Contributions: Conceptualisation: D.L.W. and P.G.H.; Designed experiment: D.L.W. and S.B.; Performed experiment and statistical analysis: S.B.; Wrote the paper: S.B.; Funding acquisition and project supervision: D.L.W. and P.G.H.; All authors revised and approved the paper.

Funding: This research was funded by BBSRC (BB.J021385/1).

Acknowledgments: We are very grateful to Dogs Trust for allowing us access and data collection to be carried out at their Centre in Ballymena, Northern Ireland. Thanks to Richard Turner for developing the smartphone application used in the study. The financial support of the BBSRC (BB.J021385/1) is acknowledged.

Conflicts of Interest: The authors declare no conflict of interest.

References

1. Stephen, J.M.; Ledger, R.A. A longitudinal evaluation of urinary cortisol in kennelled dogs, *Canis familiaris*. *Physiol. Behav.* **2006**, *87*, 911–916. [CrossRef] [PubMed]
2. Titulaer, M.; Blackwell, E.J.; Mendl, M.; Casey, R.A. Cross sectional study comparing behavioural, cognitive and physiological indicators of welfare between short and long term kennelled domestic dogs. *Appl. Anim. Behav. Sci.* **2013**, *147*, 149–158. [CrossRef]
3. Hennessy, M.B. Using hypothalamic-pituitary-adrenal measures for assessing and reducing the stress of dogs in shelters: A review. *Appl. Anim. Behav. Sci.* **2013**, *149*, 1–12. [CrossRef]
4. Stephen, J.M.; Ledger, R.A. An audit of behavioral indicators of poor welfare in kenneled dogs in the United Kingdom. *J. Appl. Anim. Welf. Sci.* **2005**, *8*, 79–96. [CrossRef] [PubMed]
5. Protopopova, A. Effects of sheltering on physiology, immune function, behavior, and the welfare of dogs. *Physiol. Behav.* **2016**, *159*, 95–103. [CrossRef] [PubMed]
6. Beerda, B.; Schilder, M.B.H.; vanHooff, J.; deVries, H.W. Manifestations of chronic and acute stress in dogs. *Appl. Anim. Behav. Sci.* **1997**, *52*, 307–319. [CrossRef]
7. Beerda, B.; Schilder, M.B.H.; Van Hooff, J.; De Vries, H.W.; Mol, J.A. Chronic stress in dogs subjected to social and spatial restriction. I. Behavioral responses. *Physiol. Behav.* **1999**, *66*, 233–242. [CrossRef]
8. Dalla Villa, P.; Barnard, S.; Di Fede, E.; Podaliri, M.; Di Nardo, A.; Siracusa, C.; Serpell, J.A. Behavioural and physiological responses of shelter dogs to long-term confinement. *Vet. Ital.* **2013**, *49*, 231–241. [CrossRef] [PubMed]
9. Hennessy, M.B.; Davis, H.N.; Williams, M.T.; Mellott, C.; Douglas, C.W. Plasma cortisol levels of dogs at a county animal shelter. *Physiol. Behav.* **1997**, *62*, 485–490. [CrossRef]
10. Koolhaas, J.M.; Korte, S.M.; De Boer, S.F.; Van Der Vegt, B.J.; Van Reenen, C.G.; Hopster, H.; De Jong, I.C.; Ruis, M.A.W.; Blokhuis, H.J. Coping styles in animals: Current status in behavior and stress-physiology. *Neurosci. Biobehav. Rev.* **1999**, *23*, 925–935. [CrossRef]
11. Sloan Wilson, D.; Clark, A.B.; Coleman, K.; Dearstyne, T. Shyness and boldness in humans and other animals. *Trends Ecol. Evol.* **1994**, *9*, 442–446. [CrossRef]
12. Horvath, Z.; Igyarto, B.-Z.; Magyar, A.; Miklosi, A. Three different coping styles in police dogs exposed to a short-term challenge. *Horm. Behav.* **2007**, *52*, 621–630. [CrossRef] [PubMed]
13. Hiby, E.F.; Rooney, N.J.; Bradshaw, J.W.S. Behavioural and physiological responses of dogs entering re-homing kennels. *Physiol. Behav.* **2006**, *89*, 385–391. [CrossRef] [PubMed]
14. Dawkins, M.S. Using behaviour to assess animal welfare. *Anim. Welf.* **2004**, *13*, S3–S7.
15. Yeates, J.W.; Main, D.C.J. Assessment of positive welfare: A review. *Vet. J.* **2008**, *175*, 293–300. [CrossRef] [PubMed]

16. Boissy, A.; Lee, C. How assessing relationships between emotions and cognition can improve farm animal welfare. *Rev. Sci. Tech. (Int. Off. Epizoot.)* **2014**, *33*, 103–110. [CrossRef]

17. Mendl, M.; Burman, O.H.P.; Parker, R.M.A.; Paul, E.S. Cognitive bias as an indicator of animal emotion and welfare: Emerging evidence and underlying mechanisms. *Appl. Anim. Behav. Sci.* **2009**, *118*, 161–181. [CrossRef]

18. Barnard, S.; Matthews, L.; Messori, S.; Podaliri-Vulpiani, M.; Ferri, N. Laterality as an indicator of emotional stress in ewes and lambs during a separation test. *Anim. Cogn.* **2016**, *19*, 207–214. [CrossRef] [PubMed]

19. Barnard, S.; Wells, D.L.; Hepper, P.G.; Milligan, A.D.S. Association between lateral bias and personality traits in the domestic dog (*Canis familiaris*). *J. Comp. Psychol.* **2017**, *131*, 246–256. [CrossRef] [PubMed]

20. Rogers, L.J. Relevance of brain and behavioural lateralization to animal welfare. *Appl. Anim. Behav. Sci.* **2010**, *127*, 1–11. [CrossRef]

21. Anderson, D.M.; Murray, L.W. Sheep laterality. *Laterality* **2013**, *18*, 179–193. [CrossRef] [PubMed]

22. Leliveld, L.M.C.; Langbein, J.; Puppe, B. The emergence of emotional lateralization: Evidence in non-human vertebrates and implications for farm animals. *Appl. Anim. Behav. Sci.* **2013**, *145*, 1–14. [CrossRef]

23. Schneider, L.A.; Delfabbro, P.H.; Burns, N.R. Temperament and lateralization in the domestic dog (*Canis familiaris*). *J. Vet. Behav. Clin. Appl. Res.* **2013**, *8*, 124–134. [CrossRef]

24. Davidson, R.J. Cerebral asymmetry, emotion, and affective style. In *Brain Asymmetry*; Hugdahl, R.J.D.K., Ed.; The MIT Press: Cambridge, MA, USA, 1995; pp. 361–387. ISBN 0-262-04144-8.

25. Rogers, L.J. Evolution of hemispheric specialization: Advantages and disadvantages. *Brain Lang.* **2000**, *73*, 236–253. [CrossRef] [PubMed]

26. Andrew, R.J.; Rogers, L.J. The nature of lateralization in tetrapods. In *Comparative Vertebrate Lateralization*; Cambridge University Press: Cambridge, UK, 2002; pp. 94–125.

27. Rogers, L.J.; Andrew, R. *Comparative Vertebrate Lateralization*; Cambridge University Press: Cambridge, UK, 2002.

28. Larose, C.; Richard-Yris, M.-A.; Hausberger, M.; Rogers, L.J. Laterality of horses associated with emotionality in novel situations. *Laterality* **2006**, *11*, 355–367. [CrossRef] [PubMed]

29. Cameron, R.; Rogers, L.J. Hand preference of the common marmoset (*Callithrix jacchus*): Problem solving and responses in a novel setting. *J. Comp. Psychol.* **1999**, *113*, 149–157. [CrossRef]

30. Reddon, A.R.; Hurd, P.L. Individual differences in cerebral lateralization are associated with shy-bold variation in the convict cichlid. *Anim. Behav.* **2009**, *77*, 189–193. [CrossRef]

31. Batt, L.S.; Batt, M.S.; Baguley, J.A.; McGreevy, P.D. The relationships between motor lateralization, salivary cortisol concentrations and behavior in dogs. *J. Vet. Behav. Clin. Appl. Res.* **2009**, *4*, 216–222. [CrossRef]

32. Siniscalchi, M.; d'Ingeo, S.; Quaranta, A. Lateralized functions in the dog brain. *Symmetry* **2017**, *9*, 71. [CrossRef]

33. Wells, D.L. Lateralised behaviour in the domestic dog, *Canis familiaris*. *Behav. Process.* **2003**, *61*, 27–35. [CrossRef]

34. Tomkins, L.M.; Williams, K.A.; Thomson, P.C.; McGreevy, P.D. Lateralization in the domestic dog (*Canis familiaris*): Relationships between structural, motor, and sensory laterality. *J. Vet. Behav. Clin. Appl. Res.* **2012**, *7*, 70–79. [CrossRef]

35. Marshall-Pescini, S.; Barnard, S.; Branson, N.J.; Valsecchi, P. The effect of preferential paw usage on dogs' (*Canis familiaris*) performance in a manipulative problem-solving task. *Behav. Process.* **2013**, *100*, 40–43. [CrossRef] [PubMed]

36. Branson, N.J.; Rogers, L.J. Relationship between paw preference strength and noise phobia in *Canis familiaris*. *J. Comp. Psychol.* **2006**, *120*, 176–183. [CrossRef] [PubMed]

37. Poyser, F.; Caldwell, C.; Cobb, M. Dog paw preference shows lability and sex differences. *Behav. Process.* **2006**, *73*, 216–221. [CrossRef] [PubMed]

38. Siniscalchi, M.; Quaranta, A.; Rogers, L.J. Hemispheric specialization in dogs for processing different acoustic stimuli. *PLoS ONE* **2008**, *3*, e3349. [CrossRef] [PubMed]

39. Mazzotti, G.A.; Boere, V. The right ear but not the left ear temperature is related to stress-induced cortisolaemia in the domestic cat (*Felis catus*). *Laterality Asymmetries Body Brain Cogn.* **2009**, *14*, 196–204. [CrossRef] [PubMed]

40. Braccini, S.N.; Caine, N.G. Hand preference predicts reactions to novel foods and predators in marmosets (*Callithrix geoffroyi*). *J. Comp. Psychol.* **2009**, *123*, 18–25. [CrossRef] [PubMed]

41. Rogers, L.J. Hand and paw preferences in relation to the lateralized brain. *Philos. Trans. R. Soc. Lond. B Biol. Sci.* **2009**, *364*, 943–954. [CrossRef] [PubMed]
42. Gordon, D.J.; Rogers, L.J. Differences in social and vocal behavior between left- and right-handed common marmosets (*Callithrix jacchus*). *J. Comp. Psychol.* **2010**, *124*, 402–411. [CrossRef] [PubMed]
43. Gordon, D.J.; Rogers, L.J. Cognitive bias, hand preference and welfare of common marmosets. *Behav. Brain Res.* **2015**, *287*, 100–108. [CrossRef] [PubMed]
44. Wells, D.L.; Hepper, P.G.; Milligan, A.D.S.; Barnard, S. Cognitive bias and paw preference in the domestic dog (*Canis familiaris*). *J. Comp. Psychol.* **2017**, *131*, 317–325. [CrossRef] [PubMed]
45. Rooney, N.J.; Gaines, S.A.; Bradshaw, J.W.S. Behavioural and glucocorticoid responses of dogs (*Canis familiaris*) to kennelling: Investigating mitigation of stress by prior habituation. *Physiol. Behav.* **2007**, *92*, 847–854. [CrossRef] [PubMed]
46. Rogers, L.J. A matter of degree: Strength of brain asymmetry and behaviour. *Symmetry* **2017**, *9*, 57. [CrossRef]
47. Westergaard, G.C.; Chavanne, T.J.; Lussier, I.D.; Houser, L.; Cleveland, A.; Suomi, S.J.; Higley, J.D. Left-handedness is Correlated with CSF Monoamine Metabolite and Plasma Cortisol Concentrations, and with Impaired Sociality, in Free-ranging Adult Male Rhesus Macaques (*Macaca mulatta*). *Laterality Asymmetries Body Brain Cogn.* **2003**, *8*, 169–187. [CrossRef] [PubMed]
48. Neveu, P.J.; Moya, S. In the mouse, the corticoid stress response depends on lateralization. *Brain Res.* **1997**, *749*, 344–346. [CrossRef]
49. Siniscalchi, M.; McFarlane, J.R.; Kauter, K.G.; Quaranta, A.; Rogers, L.J. Cortisol levels in hair reflect behavioural reactivity of dogs to acoustic stimuli. *Res. Vet. Sci.* **2013**, *94*, 49–54. [CrossRef] [PubMed]

MDPI

St. Alban-Anlage 66

4052 Basel

Switzerland

Tel. +41 61 683 77 34

Fax +41 61 302 89 18

www.mdpi.com

Symmetry Editorial Office

E-mail: symmetry@mdpi.com

www.mdpi.com/journal/symmetry

www.ingramcontent.com/pod-product-compliance
Lightning Source LLC
Chambersburg PA
CBHW051914210326
41597CB00033B/6132